ROBOTS

IN

SPACE

ROBOTS

THE SECRET LIVES OF

IN

OUR PLANETARY EXPLORERS

SPACE

DR EZZY PEARSON

The History Press

First published 2020

The History Press
97 St George's Place, Cheltenham,
Gloucestershire, GL50 3QB
www.thehistorypress.co.uk

British Library Cataloguing in Publication Data.
A catalogue record for this book is available from the British Library.

ISBN 978 0 7509 9089 9

Typesetting and origination by The History Press
Printed and bound in Great Britain by TJ Books Limited

For Mum, who would have been proud

CONTENTS

INTRODUCTION

Space. The last great bastion of exploration left to humanity. At the dawn of the twentieth century, the idea of landing on another world was a fantasy relegated to the realms of science fiction. Then, in 1957, the Soviet Union launched Sputnik 1, the first artificial satellite, and everything changed. The planets were no longer beyond the grasp of humanity. Instead, they were the next new lands waiting for those bold enough to venture out beyond the horizon.

Today, we are firmly entrenched in the Space Age. Every year, more and more spacecraft journey out into the void, heading off to another planet, moon, comet or asteroid. Each new mission is another step forward in mankind's push to explore the cosmos around us.

But for the first time in history, it's not humans that are leading the way to *terra incognita* but our mechanical envoys – robots.

These metal explorers have several advantages over us fleshy meat bags. There are places in the solar system where the radiation is so high that a human's DNA would break down within hours, but where a robot has visited. There are planets so cold that a human would freeze in moments, but where a robot has visited. There are worlds where a person would be boiled, crushed, corroded and poisoned, but where a robot has visited.

Robots can be thrown around with accelerations that would liquify squishy organs but barely even effect their circuitry. And to top it all off, you don't necessarily need to worry about bringing them back afterwards.

While robots have managed to fly past every kind of body found in our solar system, returning some incredible images, there's nothing that quite compares to those missions that have reached out and touched another world. There is something visceral about making contact with an object, even by proxy, that makes it seem more real. In today's age of photo editing, seeing is very often not believing. If you touch something, however, then you know it's real.

In my work as a space journalist,* I've been documenting from the front lines of space exploration. While there have been dozens of new robotic missions in those years, there's nothing quite like the furore surrounding those times a spacecraft touches down on another world. These are the missions that capture not just my imagination but the world's.

And yet, the history of these robotic explorers often goes unremembered, especially if accomplished by anyone other than NASA of the United States. Ask most people in the West whether or not humanity has landed on Venus and they'll probably say we haven't. In fact, the Soviets have managed to land on the hellish planet, not just once but multiple times.

Robots have been all over our solar system. They've landed on the Moon, Venus and Mars. They've dived into the atmosphere of Jupiter and driven through the tail of a comet. They've courted asteroids and returned home to tell their tale. Herein lies the heroic story of the robots that have ventured beyond the safety of our own world to reach another.

The story isn't a straightforward one to tell. It's been a complicated journey, with dozens of different nations taking part. The narrative of this book will jump around in time, instead choosing to focus on the story of each region of our solar system that we have visited.

To stop you getting lost, I'd like to clarifying a few terms. When referring to our own, I'll use a capital 'M': Moon. All other moons will have a little 'm'. I'm also going to use 'world' as a catchall term for any kind of planetary body: planets, moons, comets and asteroids.

There are also times when spacecraft have been given multiple names. For the most part, I will simply refer to them by their most common title and list any other names in a footnote.

One of the major space players today is Russia. As the former Soviet states also contributed to the nation's early space programme, I will refer to the country as the Soviet Union prior to 1992, and Russia after that.

Advances in science or exploration are never achieved solely by one middle-aged white man, despite what other history books might to try to convince you. Every single thing we know about the universe is the work

---

*   Yes. It's a real job.

of thousands of people, and it would take a book longer than this one just to list the names of everyone who deserves credit. As such, I will often refer to teams and groups.

Equally, space is no longer the domain of an elite handful of world powers. In the early days of the Space Race, it seemed that the Soviet Union and the United States were the only nations venturing into space, but in reality, dozens of countries had their eyes on the stars. However, not all of them were equally open about their exploits. In the past, the Soviet Union was notable for operating behind a veil of secrecy. Now it is China that keeps its goals close to its chest. Other nations aren't necessarily trying to be secretive but just aren't very good at blowing their own horn.

This is the reason why, in the West at least, there's often the impression that space exploration is almost solely done by NASA. While they are undoubtedly prolific, they are by no means heading out into the wilderness alone. Spaceflight may have started as two nations racing against one another, but it has now become an arena where political adversaries put aside their differences to work together towards something greater than any one nation.

That's not to say that spaceflight is some utopian ideal where everyone gets along. Politics aside, space missions are huge projects with many different teams and, as anyone who's ever worked on a group project knows, there are plenty of conflicts around every aspect of space exploration. And while space can be a grand symbol of what we can do together, it's also a great opportunity to show off on an international (and interplanetary) stage in a game of one-upmanship that has its roots in the very earliest days of the Space Race.

PART 1
THE MOON

1

DAWN OF THE SPACE RACE

Before Sputnik, the world's dreams of spaceflight were confined to the pages of science fiction and the ambitions of a few rocket engineers. In the first half of the twentieth century, groups of enthusiasts around the world created rocketry clubs, attempting to build vehicles that could one day pierce the sky. Mostly, they only succeeded in creating a lot of smoke, noise and – in the case of the student group at the California Institute of Technology who would go on to form NASA's Jet Propulsion Laboratory (JPL) – exploding one too many rockets on campus, resulting in them being banished to the nearby foothills of Pasadena.

Unfortunately, while these clubs sought to raise the human race up to the stars, the first real advancements in rocket technology would come from conflict. As Europe was gearing up for the Second World War, the work of young German rocket engineer Wernher von Braun caught the attention of Nazi military minds. Since childhood, von Braun had dreamed of exploring the stars, becoming obsessed with the rockets that might take us there. Joining the Nazi Party would allow von Braun to build his rockets, although they would be pointed at the Reich's enemies instead of the heavens. It was a compromise von Braun was willing to make. He took an SS officer's commission and set to work creating Germany's first ballistic missiles.

Initially, von Braun's efforts were condemned by Adolf Hitler as overpriced ordnance shells. Then, on 8 September 1944, the weapon's true power became apparent when a rocket launched from the Hague in Nazi-occupied Holland dropped on Staveley Road in Chiswick, west London, over 300km away, tearing it apart. Nazi High Command called it the

Vergeltungswaffe 2 (meaning retaliatory or retribution weapon). To the rest of the world, it was simply the V-2.

For months, over 3,000 V-2s pummelled Germany's enemies. But the tide of the war had already turned in favour of the Allied forces, led by the Soviet Union, the United Kingdom and the United States. By the spring of 1945 the war was all but over at Mittelwerk, the underground factory built into a hill in central Germany, where the bombs were manufactured with slave labour from the Mittelbau-Dora concentration camp. As the United States and Soviet forces closed in, the German engineers knew that their only chance to avoid being tried for their atrocities was by selling their knowledge of rocketry to the highest bidder.

Although the embers of one war were fading, the kindling of the next was already being laid down. Even though they were ostensibly allies, it was increasingly apparent that wildly differing world views would soon cause the United States and the Soviet Union to come to blows. The V-2 technology might not have been enough to save Germany, but it could shift the future balance of global military dominance.

The two powers raced each other to gather as many of the plans, hardware and personnel involved with the V-2 programme as possible. However, when forced to choose between toil in communist Russia and a comfortable life of material wealth in the United States, almost all the engineers, including von Braun, chose to go with the Americans. In return for their knowledge, the scientists would eventually be allowed US citizenship and a blind eye would be turned to their Nazi past.*

The United States had everything it needed to build an entire fleet of missiles and place itself firmly at the top of the arms race. But on 6 August 1945, the United States dropped atomic bomb 'Little Boy' on the Japanese city of Hiroshima. Its destructive power stunned the world and made the V-2 look like a child's toy. Why would the United States need the second-most terrible weapon to come out of the war when it already had the first? The German engineers were sequestered in Texas,

---

*   Despite his achievements, the ugly origins of the father of spaceflight should not
be ignored. Von Braun later claimed that joining the Nazi Party was the only way
he could continue researching rockets; he was doing his patriotic duty in a time of
war and had no idea of the horrors of the concentration camps. This is debatable,
as several eyewitnesses at Mittelwerk claimed they saw him take joy in beating the
prisoners. Although he might not have known the full extent of the atrocities, he
knew of some of them. He may have built the craft that took humanity to the stars,
but he sold his soul to do so.

where they taught the army to build copies of the now-redundant V-2 and were quickly forgotten.

With the Cold War now in full swing, the Soviet Union continued to pursue rocket technology in an effort to gain a military edge over the United States. Their efforts were led by the austere Sergei Korolev.

Korolev's road to the top had been a hard one. Before the Second World War, he'd been one of the best engineers at the Russian Jet Propulsion Research Institute but fell victim to the Great Purge, a period of government-sanctioned paranoia that lasted from 1936 to 1938. Korolev spent years in prisons and the Gulag, before serving out most of the war building rockets in a *sharashka*, a labour camp where prisoners worked in secret laboratories for the state. He was finally set free in 1944 but continued his work on rocketry.

Korolev was a difficult man, possessing a sharp temper and belligerent attitude that meant many of his contemporaries refused to work for him more than once. Yet, he was undoubtably brilliant, and his time in the Gulag had left him with a fierce determination.

His goal, like von Braun, wasn't to use rockets to blow up people but to send them into space. First, however, he would have to convince the Soviet governmental institution that oversaw many military matters, the Presidium of the Supreme Soviet, that his dreams were more than a childish fancy.

The Soviet leaders failed to share Korolev's enthusiasm for space. Korolev hoped to change their minds by revealing a prototype he'd been working on in secret – an artificial satellite. It was the first of its kind and if they'd let him fly it, Korolev could ensure the first hands to reach out towards the heavens were Soviet ones, but the leaders remained unimpressed.

Thankfully, one of Korolev's talents was playing the political game and he knew exactly how to get the Presidium on side – he told them the United States were not only building satellites, they were very close to launching one. This was, of course, a blatant ploy but it was effective. Now Korolev had not just his funding but the interest of his government.

Meanwhile, von Braun had moved to Huntsville in northern Alabama to work for the US Army and was having similar problems getting their support, while also fighting off competition from a rival rocket project being conducted by the navy. Von Braun was rapidly realising that he couldn't change military minds, but perhaps he could change civilian ones.

The science fiction of the first half of the decade had been filled with buccaneer space heroes such as Edgar Rice Burroughs' *Barsoom* novels, which transported American Civil War veteran John Carter to the Red Planet to repeatedly save the day while falling in love with Martian princesses. These works showed the public's hunger for space travel. Von Braun played on this, writing articles for magazines, using the imagery of these fictional worlds to enthuse people about spaceflight, before realising there was a better way to capture people's attention – television.

One man who was a master at using television to both entertain and inform was the animator extraordinaire, Walt Disney. At the time, he was in the process of creating several television shows to promote his new amusement park in California. Von Braun first acted as a technical consultant for several shows about space exploration but his charismatic personality and ability to simply explain complicated concepts soon put him in front of the camera. On 9 March 1955, von Braun stepped onto the screen in the first episode of *Tomorrowland*, showcasing the latest scientific advancements that could, with the right support, send people into space. Watched by 42 million people, *Tomorrowland* fuelled a new obsession with all things futuristic amongst the American public.

The final push towards space came from outside either nation, when the International Geophysical Year, a sort of scientific Olympics, laid down a challenge to the world's scientific institutions to send a spacecraft into orbit before the end of 1958. The Soviet Academy of Sciences announced its intention to join the race, despite not technically having the backing of the government that would fund it at the time.

The United States threw its hat into the ring too, but much to von Braun's consternation the government was backing the navy project, Vanguard. Although von Braun was now a celebrity and a US citizen, his project was based on stolen German technology; Vanguard had the advantage of being all-American. The army, by contrast, had been ordered to destroy all its remaining space rockets, an order the agency head, Colonel John Medaris, chose to ignore.

The Geophysical Year's challenge was won on 4 October 1957, when Korolev finally launched his satellite, now named Sputnik, from the Baikonur Cosmodrome in modern-day Kazakhstan. Mankind had made its first venture beyond the confines of Earth.

At first, neither the Soviet nor US leaders realised the momentous nature of the event, and the following day the Soviet state newspaper *Pravda* ran only a small article about the launch. However, the US media had a very different take. The 'Reds' had placed a satellite into orbit; how long before they used it to launch a bomb? Was there one already up there, waiting to turn the temperature of the war from cold to thermonuclear?

The Russian Government realised what Sputnik really was: a symbol of their prowess. Perceptions had shifted overnight from a world order in which the US military reigned, to one where Russia was the technological superpower. The Americans could no longer let this threat go unmet. The Space Race had begun.

Wanting to one-up themselves on the next space mission a month later, the Soviets hastily reworked Sputnik 2 to accommodate the first ever living creature to venture into space, a stray dog named Laika. But there was no time to work out how to safely return the poor animal. She died from overheating while still in orbit, the first casualty of the Space Race.

The mission showed that while sending a human, or indeed any living creature, into space was entirely possible, bringing them back alive was quite another thing. It would take time before the Soviets could reliably launch a human into orbit. In the meantime, the nation needed to cement its glory. They needed another headline-grabbing mission.

The most obvious target was the Moon. Even then, the Soviets realised the key to capturing the human imagination was tangibility. If they could make contact with the Moon, then they would forever be remembered as the first nation that had the audacity not just to reach the heavens but to touch them.

Touching was pretty much all the mission hoped to achieve as even a crash landing would be difficult enough. The first Luna probes were little more than balls of steel with a radio and a few instruments strapped to them. The only method of steering came from the rocket that launched them. If that rocket's aim was off during take-off, then the spacecraft would miss, which is precisely what happened to Luna 1.* Instead of crashing into the surface, it sailed past almost 6,000km away.

Rather than admit that Luna 1 hadn't gone to plan, the Soviets claimed they had intended a flyby all along, the start of a long Soviet tradition of hiding failed space missions. The Soviet Union refused to announce exactly what a mission was before they flew. If a spacecraft failed to get out of Earth orbit, then it would be given the designation Kosmos, while the propaganda arm of the Soviet Union declared it was never meant to

---

*   The spacecraft was originally dubbed 'The First Cosmic Ship' by the Soviets. In the West, it was usually referred to as Sputnik 3. In 1963 it was eventually named Luna 1. I will use the latter to avoid confusion.

go any further anyway (although the West was rarely fooled). Only once a mission was officially under way would it receive its official name and number designation. If it went wrong on the way, well then, that had always been the plan.

While this helped to give the impression that the Soviet space programme was doing far better than it was, it sometimes backfired. Instead of Luna 1 being praised, the mission was met with intense scepticism. No one knew that a lunar mission was about to take place and the probe was foolishly sent on its way on a Friday night. The majority of the mission happened over the weekend, when the telescopes that could bear witness to its flight weren't in operation.

'Everything I had seen and heard in Russia argued against the alleged fact of Lunik,' said Lloyd Mallan in *Russia and the Big Red Lie*, a book published in 1959 that was dedicated to defaming the Soviet space programme. 'The scientific community which I had studied in that enigmatic land was not capable – simply not capable – of producing any such thing… The Lunik in short, was a coolly insolent, magnificent, international hoax.'

The Soviet Union wouldn't make the same mistake with Luna 2. The moment they were sure that Luna 2 was on its way to the Moon, the Soviets made sure everyone knew where it was. The spacecraft released a cloud of sodium vapour, making it easy for telescopes to spot, but to really quiet the naysayers they would need a more accurate measurement.

One facility that could help them was the huge radio telescope at Jodrell Bank, located just outside Manchester in north-west England. The telescope was making a name for itself tracking spacecraft and if Jodrell Bank said that the Soviets had crashed into the Moon, then they had.*

Back in the Soviet Union, Korolev and his team waited in the control room as their spacecraft drew ever closer to the Moon. It was dead on target, heading for an area known as the Marsh of Decay. The spacecraft continued to transmit when, at 21:02 Universal Standard Time (UST) on 13 September 1959 the signal suddenly fell silent. The probe had impacted the Moon at an astonishing 10,000km/h.

---

* The Soviets appeared to have a second case of bad timing. The collision once again occurred over the weekend, and Jodrell Bank director Bernard Lowell almost missed the event as he had far more important matters to attend to – his weekly cricket match. It was only when he received a direct call from Moscow that Lowell realised the significance and ditched his cricket whites to get back to the office.

Mankind had reached out and touched another world, and it was a Communist hand that had done so. The day after the landing, Soviet leader Nikita Khrushchev met with US President Dwight D. Eisenhower during a rare trip to the United States. Ostensibly as a gesture of goodwill, he presented the US leader with a replica of the Luna 2 lander, complete with the dozens of pennants bearing the hammer, sickle and star it was meant to scatter across the lunar surface. In truth, the spacecraft had been a 'hard-lander' – meaning it crashed into the surface at speed, rather than a controlled descent – and Luna 2 most likely vaporised on impact. But the message was clear. The path to the Moon was being forged by the Communists.

The gift was a power play, one Eisenhower refused to be drawn into. While there were many in the US government, military, media and public who clamoured for a hasty retort to the 'Red Threat', no matter the cost, the president knew the rewards of a space programme were best reaped slowly. Rather than racing towards a short-term game of one-upmanship, President Eisenhower called for a long-term plan of robotic missions that would help set the United States up for the future.

It was an admirable goal, but the US space programme was already floundering. The US Navy launched Vanguard, their response to Sputnik, on 6 December 1957, but it made it little over a metre off the ground before crashing back to Earth in front of the world's media. Dubbed 'Kaputnik' (*Daily Express*), 'Flopnik' (*Daily Herald*) and 'Stayputnik' (*New Chronicle*), the United States' failed effort left the nation humiliated.

With the navy unsuccessful, Colonel Medaris stepped forward with the army's offering. He had risked being court-martialled to continue building his and von Braun's satellites in secrecy, but it paid off. On 31 January 1958, the United States launched its first successful satellite, Explorer 1. Colonel Medaris was America's knight in shining armour.

The Vanguard/Explorer face-off had highlighted that the current system for developing US space technology wasn't working. Each branch of the military was fighting for control, working against each other when they should have been collaborating. They needed a single body to bring together all aspects of the space programme and on 29 July 1958, Eisenhower signed into creation the National Aeronautics and Space Administration, NASA.

It was a smart move, one that Korolev suggested the Soviets replicate, but his request was ignored by Khrushchev. Instead, the Soviet space programme remained in the hands of several different design bureaus, many of them working for completely different ministries. It would prove a costly mistake.

Despite these organisational problems, the Soviet space programme still delivered incredible new missions with a regularity that astounded the world and annoyed the United States. On 4 October 1959, a third lunar probe, Luna 3, launched from Baikonur.

Unlike Luna 1 and 2, which had run on batteries charged before launch, Luna 3 had solar panels. Solar cells, which convert the energy from sunlight into electrical current, had been around since 1883, but proved too unwieldy for industrial use. Seventy years later, in the 1950s, Bell Laboratories in the US state of New Jersey succeeded in making them more efficient, only to make them completely unaffordable as well. While they were no good for large-scale use, in situations where the high cost wasn't an immediate barrier, such as the well-funded space programme, solar panels could be used as portable power stations.

Now that it had solar panels, Luna 3 wasn't confined to the few hours of battery power its predecessors had been. Spacecraft would now be able to operate for months, perhaps even years. It was a major evolution in the technology of spaceflight.

Luna 3's solar panels generated enough power to operate a camera. On 7 October 1958, Luna 3's sensors picked up the bright light of the Moon and set the camera running. They were still going as Luna 3 missed the surface, this time on purpose, and swung around into orbit. Once done, the film was transferred to an on-board lab to be chemically developed, dried, scanned in and sent back to Earth.

The images were fuzzy, but contained a view never seen by human eyes – the lunar far side.

Wanting to ensure that everyone knew about their triumph, the Soviets sent a model of Luna 3 on tour. Among those who came to see it were several CIA operatives, who were shocked to discover the spacecraft wasn't just a model but a flight-ready spacecraft. If they could just get a closer look, they might be able to find out how the Russians had reached the Moon.

One bribed truck driver later, the CIA spies were secretly dismantling one of the technological marvels of the age. After taking dozens of pictures and measurements of its inner workings, the team put it all back together and returned it to the driver. No one was any the wiser to the lunar space-craft's terrestrial detour.*

---

*   As it remained classified until 1995, exactly how useful this stolen intelligence was remains uncertain. It is, however, a fantastic story and I'm at a loss to explain why it has yet to be made into a fabulous spy-caper mini-series.

This level of espionage, risking an international incident to steal the secrets of Soviet spacecraft, shows exactly how desperate the United States was to catch up. The triumph of the Luna missions wasn't just seen as a success for the Soviet space programme, but as a sign of the superiority of the communist way of life. Nations like France and the UK were beginning to lose faith in the United States' military superiority and looking towards the Soviet Union, a situation the Americans found untenable. The United States needed to save face, which meant getting to the Moon. And fast.

2

RANGER DANGER

NASA's answer to the rapidly progressing Soviet Luna programme was Ranger, a thirty-six-month-long venture to send a series of spacecraft to smash into the Moon. This was a very quick turnaround for a programme consisting of all-new technology. The trade-off for speed, however, was a much higher chance of failure than anyone at NASA would have liked, but it was the only way to keep pace with the Soviets.

There were initially five missions planned, all built around the same generic form: a conical body, with two solar panels and a dish antenna sticking out from the base. The idea was to use an incremental approach, at first staging simple missions to get the basics right before advancing to a more complex spacecraft. The first two missions, dubbed Block I, would test the systems needed to launch towards the Moon, but wouldn't attempt to go near it. That would be left to the Block II probes, which would be impactors launched on a collision course with the Moon.

As NASA couldn't be the first to touch the Moon, they concentrated on making a more impressive landing. Like Luna 2, the mission would be a hard-lander, but the Ranger spacecraft would have a package of scientific instruments that Luna 2 had lacked. Although this wouldn't survive, the spacecraft would take photographs as well as magnetic and particle measurements up to the final moment before impact.

There were even plans to carry a small seismometer that would eject just before impact. It would balance on the nose of the spacecraft's cone, surrounded by a balsa wood ball, which would crush on landing to absorb the impact. One of the biggest questions around the Moon at the time was whether its craters were made by volcanoes or meteor impacts. If lunar volcanoes were responsible, then the Moon could still be geologically active, with insides that were partially liquid and moving around. The seismometers would be able to detect these motions as moonquakes.

Unfortunately, Ranger was beset by problems from the start. NASA had only just formed, and new departments were joining every day. No one was sure exactly who was in charge of what. The Jet Propulsion Laboratory (JPL), now part of NASA, was the contract manager in charge of designing and building the spacecraft, but the air force was in charge of the launch vehicle and von Braun's army team were also involved. Throughout, the military-based wings of the mission were still squabbling over who should be building rockets, and frequently refused to talk to each other.

To make things even harder, NASA decided that Ranger would be far more complicated than it strictly needed to be. To keep costs down in the long run, the agency planned to create a spacecraft design that could be used to visit not just the Moon, but other worlds as well.

Shortly into the project another, more tangible, problem arose with Ranger's rockets. When it comes to space missions, weight − or more accurately, mass − is king. To get something off Earth, you have to point it upwards and give it enough speed to escape Earth's gravity (around 11.2km/s, thirty-three times the speed of sound). Giving something that amount of speed takes a lot of energy, which is why you use a big tube filled with explosives, otherwise known as a rocket. However, the heavier the object you're trying to launch − known as the payload − the more energy it takes and the more fuel you need. This makes your rocket heavier, which means you need more fuel, and so on.

Whereas the Luna probes had been sent directly from Earth to the Moon, Ranger would be launched using an Atlas rocket into what's known as a 'parking orbit' around Earth. From here, a second smaller engine, the Agena, would send the spacecraft on towards the Moon. Doing so would widen the potential launch window − the range of time where the positions of the Moon and Earth make the journey possible.

The Atlas–Agena rockets were being built for the air force by a private aerospace company called Lockheed. On 11 July 1960, the company submitted a set of figures stating it would be able to carry a payload 34kg less than NASA had been expecting. This was fine for the Block I flybys, but the Block II landers would have to shed serious weight. No one was sure how the discrepancy could have happened, or if it was even accurate, but there was no time to investigate fully. They'd just have to lose the weight.

At first the team tried to 'nickel-and-dime' their way out – shaving down insulation, using thinner wires, shaving down struts and even drilling holes in the sides of internal panels until some began to resemble Swiss cheese. But it wasn't enough. Something big was going to have to go.

Unwilling to sacrifice the spacecraft's main systems, NASA elected to throw out the redundant ones. If something breaks on a spacecraft, you can't send out a mechanic to fix it, so to prevent an entire mission being lost due to a broken radio or faulty clock, most spacecraft have redundant systems built in so that they'll carry on operating even if one part fails. Without these redundancies, one broken wire could cause the entire mission to fail. It was a big gamble, but there was no choice.

To add insult to injury, a few months later Lockheed reported back that not only could they actually launch the originally specified weight, they could handle more. However, Ranger was already running late. If the project team waited around to refit the gutted spacecraft, they'd miss the thirty-six-month deadline they'd been set. The first Rangers were launched as they'd been built.

The Block I Ranger missions, the flybys, began in August 1961. Ranger 1 took a while to even get off the ground, with launch delays caused by everything from power interruptions to accidentally deploying the solar panels while still on the launch pad. When it finally launched, the rocket's upper stage – the part meant to take it to the Moon after the first stages got it into space – failed, stranding it in Earth orbit. Ranger 2 didn't fare much better four months later.

Both these issues were with the rockets, not the spacecraft. The Rangers had performed exactly as expected and the schedule was advancing, so NASA decided to move onto Block II, and start with the landing missions.

The more time went on, the more problems piled up. In May 1961, the company responsible for building the balsa wood-covered seismometers, Aeronutronic, reported their test drops of the landers in the Mojave Desert were going badly. Many simply didn't survive and those that did suffered electronics issues. The fault seemed to lie with the stringent sterilisation procedures that NASA had imposed on the spacecraft.

While most spacecraft fly through the void of space, Ranger would be touching down on a completely alien world. While more fanciful ideas of hidden lunar civilisations had been disproved, there was still a chance the Moon could be home to bacteria-like life.

Knowing how aggressive Earth microbes can be once they get a foothold, Nobel Prize-winning geneticist Joshua Lederberg feared that an

Earth probe could carry bacteria to the Moon and other planets where they would spread, potentially wiping out any existing forms of life. Lederberg wrote:

> History shows how the exploitation of newly found resources has enriched human experience; equally often we have seen great waste and needless misery follow from the thoughtless spread of disease and other ecological disturbances. The overgrowth of terrestrial bacteria on Mars would destroy an inestimably valuable opportunity of understanding our own living nature.*

Following Sputnik, Lederberg campaigned for extensive planetary-protection measures: every spacecraft that would touch another world must be decontaminated before leaving Earth. His pleas were heard and in October 1959, Abe Silverstein, the NASA Director of Space Flight, ordered all impactors to be sterilised.

Doing so, however, is easier said than done. Not only is life everywhere on Earth, it's extremely tenacious. Completely eradicating bacteria, and keeping them eradicated, is impossible, although you can get damn close. The best the Ranger team could do was bake the spacecraft at 125°C for 24 hours and then wash it with toxic ethylene oxide just before launch. The temperatures would damage some of the more delicate components, but the Ranger team thought that with just a few reworkings the spacecraft would be able to survive the procedure.

They thought wrong. The sterilisation ended up destroying several key components, leaving a non-functioning spacecraft. NASA waived some of the more extreme protection measures for the more fragile parts, but as Block II progressed, more component failures were discovered. The team began to consider Ranger 3 even reaching the Moon a success. If the lander actually managed to set down or send data, it would be a bonus. Besides, Rangers 4 and 5 were always there for another run.

But, while NASA was worrying about frail robotics, they failed to account for the biggest problem in any space programme – the fallibility of the humans operating them. Ranger 3 launched on 26 January 1962 and things seemed to be well under way, until it came time for the mid-course correction. Ranger had thrusters that could subtly nudge the

---

*   J. Lederberg, 'Exobiology: Approaches to Life Beyond Earth', *Science* (1960).

spacecraft to make sure it was on target. When the code had been written for Ranger 3's mid-course correction, an error meant the thruster's direction was input backwards. The engine fired the wrong way and Ranger 3 missed the Moon by over 36,000km.

This error was human, not robotic. The team should be able to do better next time.

They didn't.

The rockets delivered the next spacecraft into orbit as expected, but when the spacecraft phoned home, there was no telemetry – the data that tells NASA where the spacecraft is and how it's doing. One of Ranger 4's clocks had stopped ticking. It couldn't keep track of its position and it was tumbling as it flew towards the Moon. The spacecraft transmitted for 64 hours, before falling silent on impact with the Moon. US hardware had finally made it to the lunar surface, but it was a hollow victory.

Four missions in and Ranger still hadn't done any science. The Ranger mission was not getting off to an auspicious start. However, it was about to meet the biggest obstacle to the scientific exploration of the Moon: John F. Kennedy (JFK).

In the early 1960s, both the United States' and Soviet plans for the exploration of the Moon and the planets were ostensibly scientific, but there was no denying that the missions were being used as political weapons. The Soviet space programmes posed an unstated threat to the United States: if we can launch a satellite and land it on the Moon, we can launch a bomb and drop it on Washington DC. The United States' failure to keep up was not just a sign of their science programme's downfall, but a sign of their military prowess lagging behind the Soviets.

The 'missile gap' – the perceived disparity between the few missiles the United States had built compared to the many the Soviets were assumed to have – was one of the key topics of the 1960 US presidential election, one that JFK made a key campaign policy. The current president, Eisenhower, had refused to be drawn into the Space Race, favouring long-term progress over short-term gain. Despite having been against the space programme while he was a senator, JFK now painted Eisenhower's lacklustre approach to space as a failure to respond to the Soviet threat, hoping to undermine his political opponent. It worked, and on 20 January 1961, JFK became the 35th President of the United States.

Just a few months later, the Soviet space programme achieved its most headline-grabbing milestone yet when on 12 April 1961, Yuri Gagarin became the first human to orbit Earth.* With every success, the Soviets cemented their image as the superior power and, over time, other nations of the world were beginning to lend their support to the communist state over the United States.

The United States needed a win. A big one. Some grand moment that would forever cement it as the leader of spaceflight.

NASA's own human spaceflight programme was already in motion. The idea of a human lunar landing had been kicking around since 1960, but Eisenhower had baulked at the astronomical cost. Now, JFK saw a chance for the United States to take back the stars. A moon landing was something he could get the nation behind, something he could use to inspire and, importantly, had a single goal that he could point to and say, 'Look what we achieved'.

On 25 May 1961, Kennedy stood before Congress and announced that the United States should 'commit itself to achieving the goal, before this decade is out, of landing a man on the Moon and returning him safely to the Earth'. This was the speech that would launch the Apollo programme.

It's undeniable that if the president wanted to create a single, defining moment, Apollo was it. Half a century later, Apollo 11 is still used as the symbol of what humanity can achieve when we set our minds to it. What Apollo was not, and was never meant to be, was a scientific enterprise.

As soon as Apollo was given the go-ahead, it became NASA's focal point. Unfortunately, it came at the expense of almost every other project. The existing lunar programmes were gutted. Their scientific content was wrenched out, to be replaced with tasks that would serve Apollo. Planned planetary missions to Venus and Mars were in jeopardy of being cancelled outright. Faced with an ever-ballooning budget, US Congress – the entity that holds NASA's purse strings – threatened to cut the funding for any-thing that wasn't related to Apollo.

James Webb – the visionary administrator who saw the agency through its formative years – knew that the space programme was more than a mere political tool and fought the move. 'So far as I'm concerned, I'm not

---

*   In a twist of fate, Alan Shepard, the first US citizen to reach space, could have launched before Gagarin but von Braun insisted on one last uncrewed test flight, delaying his launch. If Shepard had beaten Gagarin, there's every possibility that JFK would not have felt the need to commit to Apollo. How different might the story of space exploration be if he had?

going to run a program that's just a one-shot program,' Webb said. 'If you want me to be the administrator, it's going to be a balanced program that does the job for the country ...'*

NASA kept its science funding but Webb couldn't save everything. In October 1962, the Director of the Lunar Planetary Programme, Oran Nicks, was tasked with working out which of NASA's robotic missions could be used to support Apollo.

Ranger was the obvious choice. The probes could be sent to scout out potential landing sites, photographing the terrain, while the results from their collisions could answer one of the most pressing questions of the project – could the lunar surface even support the weight of a human?

The programme was given funding for four new Block III missions that were designed to pave the way for Apollo. Despite the scientific scope of these probes being cut back, the Ranger team welcomed the announce-ment of four more chances to land on the Moon, especially when on 18 October 1962 Ranger 5 lost power shortly after reaching orbit. The culprit was again found to be the sterilising heat treatment of the compo-nents causing a short circuit.

The many failures of Ranger had not gone unnoticed. The programme was subjected to an extensive review, with the board of inquiry ultimately decreeing that the harsh sterilisation procedures were to blame. As the Moon was almost certainly barren of life, the planetary protection meas-ures were deemed unnecessary and dropped. To improve matters further, the probes would no longer be designed as universal spacecraft and the redundant systems were put back in.

Now Ranger was the top priority on NASA's flight schedule; the team just had to build a Block III spacecraft that would achieve its new goals. The seismometer was the first to go, and the rest of the science instru-ments were replaced by television cameras, which digitally recorded and transmitted the image. This was vital as the spacecraft was destined to smash itself against the surface, rendering it unable to develop the images on board as Luna 3 had done.

*   Transcript, James E. Webb oral history Interview 1, 4/29/1969 by T.H. Baker, Internet copy, Lyndon Baines Johnson (LBJ) Library, University of Texas.

When Ranger 6 launched on 30 January 1964, it was heading towards one of the dark patches of the Moon, known as the 'lunar maria'. These regions make up the most obvious features on the Moon when viewed from Earth and are thought to be plains flooded by now solidified lava. Ranger 6 was heading towards Mare Tranquillitatis, the Sea of Tranquillity.* Like all the new Ranger missions, its destination was a potential Apollo landing site.

The launch proceeded completely to plan, barring a momentary hiccup when the TV system turned on for just over 1 minute. As the spacecraft successfully redirected itself for lunar impact, JPL's director, William H. Pickering announced to the press that he was 'cautiously optimistic' for the mission's success.

Eighteen minutes before impact, the TV cameras were due to begin warming up for the big show. Nothing happened. Thirteen minutes until impact, the cameras were due to switch to full power. Nothing happened. Ten minutes out, Ranger 6 was supposed to start taking images. Again, nothing happened.

At 09:24 UST, the control team were supposed to be receiving the last images from the spacecraft before it impacted the surface. Instead, all they had to watch was the radio signal from the spacecraft as it suddenly went dead. It seemed the camera's early switch-on had been caused by an electrical short that destroyed its power supply, killing the television system. Ranger 6 had made impact with the Moon, but it had done so without taking a single photograph.

Ranger seemed cursed. When Ranger 7 launched on 28 July 1964, heading towards Mare Nubium, the Sea of Clouds, the press room was packed with journalists waiting to hear how the latest Ranger would meet its untimely end. But at the 18-minute mark, the relieved team declared the cameras were warming up. The announcement they were at full power was greeted with a round of applause.

At 13:08 UST, Ranger finally succeeded in taking an image of the lunar surface. The video stream was coming in just as it should be. Then, at 13:25 UST, Ranger 7 slammed into the surface. Finally, NASA had done

---

*   The large features of the lunar landscape received their names in 1651, when astronomers Francesco Grimaldi and Giovanni Battista Riccioli created one of the first lunar atlases.

it. The United States had not just landed on the Moon but had broadcast themselves doing it.*

Ranger 7 took 4,316 pictures during its descent. The closest images had resolutions as fine as half a metre – enough detail to hunt out the Apollo landing sites. Eager to capitalise on their success, the first press conference was held an hour after the impact. That evening, scientists broadcast their findings live on national TV, as the press clamoured for an instant scientific interpretation of images taken just hours before.

The big question was whether the Moon could support a craft large enough to hold humans. The pictures had shown several large rocks on the surface – if the surface could support a boulder it could probably support a human.

Despite the mission's goals being Apollo based, the images still fed a scientific community that had been waiting for years. For the first time, geologists could count the number of meteor craters on the lunar surface. The longer a landscape has been around, the more times it will have been scarred by space rocks and so the number of craters can be used to date how long it's been since the world's surface has been refreshed by lava. The Moon appeared to have craters down to the resolution of the closest images, meaning the surface was at least 3.6 billion years old.

This constant rain of meteorites has smoothed away the edges of mountains and craters, leaving a softly undulating surface. The craters appeared to have been filled in by ancient lava flows that had now cooled, but without the seismometer there was no way to tell if molten lava still moved beneath the surface. Nor could they tell whether the lava had come from inside the planet or had been created by the heat of the impact itself.

With Ranger 7 such a success, the controllers were eager for Ranger 8.** The mission launched successfully on 17 February 1965 and made impact with the Sea of Tranquillity on 20 February. As the spacecraft

---

*   Shortly after the landing, a case of champagne appeared in the control room and the NASA staff were treated to a well-earned drink. Presumably, no one had felt like celebrating after the previous six failures.

**   One superstitious soul put the success of Ranger 7 down to the fact that someone had brought peanuts into the control room on launch day. Hoping to ensure the success of Ranger 8, they brought in peanuts again as a lucky charm. The tradition of bringing peanuts to planetary launches survives to this day.

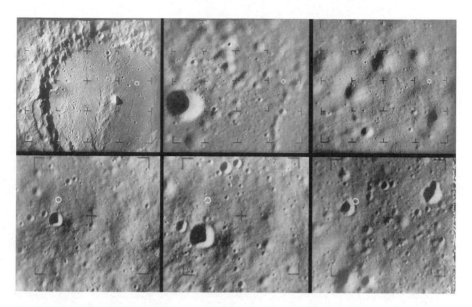

A sequence of images taken by Ranger 9 as it approached the lunar surface. The image taken furthest out is in the top left and the nearest in the bottom right. The eventual crash site is marked in all the pictures by a white circle. (NASA/ JPL-Caltech)

came in at a different angle, it was able to take twice as many photos as Ranger 7 – 7,137 in total. Unfortunately, it also meant the spacecraft was moving faster, smearing the last few frames and the best resolution was only 1.5m.

This site seemed much rockier, and the images supplied a good view of craters that looked, at first, as if they might be a caldera, the crater formed by a now long-dead volcano collapsing in on itself. A closer examination revealed they were actually created by meteor impacts.

Ranger 7 and 8 seemed to show the craters were largely the same. So it was decided the final mission, Ranger 9, would take a look at something completely different. The Ranger landing team managed to find a scientifically interesting location that still appeased the Apollo people – Alphonsus, a crater that was believed to be volcanic in origin.

When Ranger 9 crashed into the Moon on 24 March 1965, it performed a 'terminal manoeuvre', pointing the camera down the direction of its motion to prevent blurring. The final resolution was just 25cm, an incredible achievement.

After such a troubled beginning, it was a magnificent end to Ranger. The project had been intended to simply keep up with the Soviets, yet by its end had outpaced the United States' rivals by three impacts to one. After being hijacked by Apollo, Ranger had served its new masters well, returning close-up images of the surface – a feat the Soviets had failed to achieve – and given tentative evidence that the surface could support a heavy lander.

There were still many questions the Apollo teams needed answers to before they would risk American lives. Those would be up to another mission to answer. Ranger was over.

3

THE SOVIETS RETURN
TO THE MOON

After Luna 3, the Soviets' path back to the Moon was a little more circuitous than their US counterparts. The nation had largely abandoned the Moon in favour of the planets, while its human spaceflight efforts remained in low Earth orbit. For many years, the Soviets had assumed that the US moonshot was a case of fine rhetoric on JFK's part rather than a serious commitment to lunar exploration. But as Congress poured more and more money into the project, it became apparent that the Americans meant to go to the Moon, and they would get there before the Soviets.

Korolev refused to let this stand. If the United States reached the Moon first, it would outshine all the achievements the Soviets had made before this point. Starting in autumn 1963, Korolev pushed Khrushchev, declaring that not pursuing a Soviet human lunar landing would be a mistake, if not downright unpatriotic.

In summer 1964, the Soviet Union committed to sending a human to the Moon but did so in secret. While the West believed the Soviet space programme was going from strength to strength, the truth was very different. Behind the Soviet Union's veil of secrecy lay a pile of broken probes and exploded rockets, their failures far outstripping their successes. There was a serious risk they would fail to land a human on the Moon at all. While beating the United States to the surface without warning would be a major coup, announcing they were making the attempt and then failing would be a disaster.

The Soviet space programme abandoned the exploration plans they'd been working on since Luna 2 – a long-term investigation into human spaceflight while exploring Mars and Venus robotically – and refocused on the Moon. It was now the Soviets' turn to plunder their existing space missions in service of a crewed lunar landing.

The Soviets had been planning on exploring Mars and Venus robotically with a spacecraft called the Ye-6. These were designed at Lavochkin, a secret aerospace facility in the Moscow region. They would be 'soft' landers, meaning they would use retrorockets – small thrusters that fired in the opposite direction to the spacecraft's travel, slowing them down during approach to the surface. Landing at a slower speed meant the Ye-6, and the scientific instruments on board, could survive the landing.

The question was how long exactly to burn the thrusters for. This was before the days of on-board computers. All the burn times had to be worked out before the spacecraft flew and encoded with mechanical timers. If the thrusters didn't burn hard enough, they wouldn't shave off enough speed and the spacecraft would crash. Equally, if they burned too hard, they'd run out of fuel before reaching the surface and the space-craft would begin to fall under the Moon's gravity, picking up speed until they were travelling too fast to survive the landing. The final decision was to turn them on 75km from the Moon, reducing the speed from over 2.5km/s to almost nothing in around 46 seconds.

As a final precaution in case the calculations were out, the spacecraft had a boom sticking out of the top that could sense when it had struck the lunar surface. When it did, it would deploy an airbag to absorb the shock of a potentially bumpy landing. The actual lander itself was suitably sci-fi – a smooth steel egg just over 1m high that unfolded its four 'petals' to stabilise the craft and reveal a package of instruments within.

The Soviets made several attempts to launch their Ye-6 lander, but failed to reach Earth orbit at an alarming rate.* In fact, the rate was so high that when the United States published a list of the failures it had been able to spy out, many people did not believe the numbers: surely no nation could incur such losses and continue with their endeavour?

---

*   I say several because Soviet secrecy means we don't have a very good idea of what the actual numbers are. The records for much of the early Space Race didn't become apparent until after the fall of the Soviet Union in 1992, and by that point many of the records had been misfiled, destroyed by mould or burned as fuel during a blackout.

Their unlucky streak finally ended on 2 April 1963 when Luna 4 managed to get out of Earth orbit, and on towards the Moon. The mission was announced to the world with only the vague description of travelling 'to the vicinity of the Moon', with no indication of its real purpose.

However, its success was short-lived. Luna 4 soon lost its astronavigation – the systems that detect the positions of bright stars and navigate by them as mariners have done for centuries. Without this, the spacecraft had no way to correct its course. It would miss the Moon. The Soviets once again claimed they'd always intended a flyby, but few in the West believed them.

An investigation into the Ye-6 turned up a multitude of problems. These seemed to have one root cause: the mission had been rushed and quality control had suffered as a result.

With the issues fixed, hopes were higher for Luna 5 when it launched on 3 May 1965. It managed to make impact with the Moon but mostly by luck rather than design. Ground Control lost control shortly into the flight, meaning none of the in-flight thruster burns meant to refine its trajectory happened. The initial direction when leaving parking orbit had been good enough to at least hit the lunar surface, but the impact was considerably more terminal than the controllers were aiming for.

On 8 June 1965, Luna 6 suffered the opposite problem to its predecessor. While Luna 5's thrusters wouldn't turn on, Luna 6's wouldn't turn off and it ended up missing the Moon by over 160,000km. Ground Control still sent the command for the probe to proceed with its landing procedure. It did so perfectly, albeit in the void of space, rather than on the Moon.

In October, the Soviet Union celebrated the eighth anniversary of Sputnik as they prepared to launch Luna 7. Elsewhere in the lunar programme, the orbiter missions were returning fantastic images of the far side of the moon. Surely the auspicious date was a good omen for Luna 7?

The hope seemed well founded, with Luna 7 sailing through all its initial stages flawlessly. At 8,500km from the Moon, Luna 7 prepared to fire its thrusters before landing when its astronavigation systems also failed. With no way to orientate itself, the spacecraft didn't know which direction to fire the thrusters and it crashed, just as Luna 5 had done. Yet another failed mission.

For Luna 8, the stakes were raised even before it reached the launch pad on 3 December 1965. Soviet leader Khrushchev, who had always been a champion of the Space Race, had been deposed a year earlier and replaced by Leonid Brezhnev. The new leader was displeased with the seemingly endless string of failures coming out of the Luna programme. He had had enough. If the lunar programme couldn't deliver a success, then it would be cancelled.

Everything went well. The mission launched, left parking orbit and performed its mid-course corrections. As it came in towards the lunar surface, it deployed its airbags ahead of the final deceleration burn. Just a few more seconds …

Suddenly, the spacecraft went spinning. A piece of metal had snagged the airbag, and the gushing air knocked it off course. The spacecraft, sensing it wasn't aligned with the Moon, only fired its thrusters for 9 seconds during the few moments its spin pointed it towards the Moon – nowhere near the 46 seconds it needed to slow down – and it crashed. It was the programme's eighth failure in a row.

Godfather of the Soviet space programme, Sergei Korolev, used every connection he had to keep the Luna missions alive. He was given one more chance. One more mission, and if it failed then this truly would be it.

Tensions were high at Mission Control as Luna 9 was sent on its way on 31 January 1966. With the previous missions failing well into the flight, no one was willing to get too hopeful when the first stages went off perfectly. The mid-course correction went to plan. The retrorockets fired, and this time the airbags deployed without puncturing themselves.

Moments before the descent craft crashed into the lunar surface, the 99kg lander successfully jettisoned. The radio the team had been using to contact Luna 9 then died with the main spacecraft, forcing them to wait to receive the lander's own signal. Four minutes later, it started broadcasting.

The nation rejoiced at the sound of those familiar beeps and hums, showing that the Soviet Union had led the way to another milestone, the first soft landing on another world. As the minutes passed, the first images from the lunar surface began to trickle home. Wanting to maximise the publicity and propaganda potential of these images, the Soviets didn't release them straight away.

However, the Soviets weren't the only ones who were listening to Luna 9. Jodrell Bank was eavesdropping on the probe, and one of the technicians recognised the signal. Having previously worked in a newsroom, he realised the format was the same as that used to transmit newspaper images across the globe – Radiofax. When Terry Pattinson, a reporter for the *Daily Express* newspaper based in nearby Manchester, heard this, he loaded the paper's Radiofax machine into a lorry and drove it over to the observatory.

'The *Express* got the photos before the Russians and we sold them to the rest of the world,' Pattinson recalled in the paper on 19 July 2019. 'My editor turned to me and said it was the scoop of the century.'*

The picture Pattinson developed was a lunar panorama. The surface appeared to be covered in dust the texture of snow with a few ragged pebbles. As the Sun changed position in the sky during the following days, its angle gave a better contrast on the rock, revealing the surrounding landscape. However, the most important insight about the surface was that it was strong enough to support the weight of the probe, meaning it would probably support a human.

The only other scientific instrument on board was a radiation detector, which measured just 30 millirads,** far below what is was dangerous to humans. Future moonwalkers would encounter many hazards on the Moon, but radiation sickness wouldn't be one of them.

The probe lasted for three days on the surface before ceasing its broadcast. But as the world celebrated the news, Korolev, who had worked so hard to put Luna 9 there in the first place, could not join in. He'd died just two weeks before. Plagued by health problems from his years in the Gulag, a routine surgery had proved fatal. The Russian space programme had lost its guiding light.

Korolev might not have been the best engineer or manager, but he was a great leader. Against the constantly shifting landscape of Cold War Soviet politics, he kept the plates of the space programme spinning. His successor, Vasili Mishin, was an excellent engineer who had worked alongside

---

*   It's uncertain whether this was a genuine oversight, or if the images were purposefully left unencrypted by the spacecraft designer to ensure they were released to the world, rather than being jealously guarded by the Soviet media mafia. We do know that subsequent missions coded their transmissions to stop them being stolen as well.

**   To put this in perspective, this is a bit less radiation than the amount you receive from the potassium in your own body every year.

Korolev since 1945, but he was neither a leader nor a political operator. The Soviet space programme would never be the same again. Although the work of their robotic landers showed that it could be possible to land a human on the surface of the Moon, any real hope of the Soviets succeeding in doing so died with Korolev.

4

SCRATCHING AT THE SURFACE

Ignorant of the troubles going on behind the scenes of the Soviet space programme, NASA was hit hard by the news of Luna 9's landing. The Soviets had beaten them to another lunar milestone. America was losing the Space Race.

If the United States could not be first to the Moon, they would just have to do a better job once they got there. Ranger was only ever meant to be a stopgap, buying NASA time to develop a more refined soft-lander – the Surveyor programme. Although they were originally intended as scientific missions, the Surveyors would now gauge the lunar surface's suitability for a human landing, setting the stage for Apollo.

Standing 3m tall with legs 4.3m long, the probes towered over the humans building them. Each had three legs arranged in a tripod, a configuration that would provide a stable footing on uneven ground. Most of the instruments were arranged around the central body at the crux of these legs, although a single mast stuck out of the top, carrying the solar panel and antenna. They were a huge creation, but their slimline struts and white paint jobs made the Surveyors look flimsy rather than intimidating.

With Ranger's three-year programme finished, the new Surveyors were meant to step in, but testing uncovered several serious defects. The probes were finally deemed flyable in January 1966, only to be beaten to the surface by Luna 9 just a few weeks later.

It was a demoralising near miss. But while Luna 9 might have got there first, there was no denying that the Surveyors were the superior machines. Like the Luna landers, they used retrorockets to slow down from over 9,500km/h to around 400km/h. Surveyor had no airbags. Instead, once the main engines ran out, chemical thrusters, known as vernier propulsion, would guide Surveyor down onto its feet. These would be designed to absorb what was left of the spacecraft's velocity, reducing the chance of the spacecraft bouncing off the surface.

The first mission would concentrate on landing on the surface in one piece, taking images from the surface and practising the logistics of communicating home using the Deep Space Network – the series of radio outposts across the world that NASA set up in 1958 that is still in use today. If all went to plan, then the spacecraft should be able to survive for months, not just the few days Luna 9 had managed.

These missions were very much in service of Apollo. They focused on learning how to land a large spacecraft on the Moon and assessing the surface. Any geological investigation done along the way would be entirely incidental. The Surveyors would all touch down in potential Apollo landing sites, which had themselves been picked from the extensive maps created by the Lunar Orbiter Programme.* Surveyors would serve as scouts, assessing the lay of the land before any human life was risked. The first destination would be Oceanus Procellarum, the Ocean of Storms, the eventual landing site of Apollo 12.

The timing of the missions was carefully chosen. Despite being in the cold vacuum of space, a spacecraft can overheat by sitting in the full glare of the Sun. The team wanted to ensure Surveyor landed at the beginning of the lunar day (twenty-nine and a half Earth days). This would give the landers' cameras enough light to see by while allowing enough time before the heat of lunar high noon, which would force the mission to take a siesta for a few Earth days until things cooled off again.

For Surveyor 1, that narrow launch window would begin on 30 May 1966. Three months after Luna 9, the United States fired their answering salvo. After the raft of failures from both the Ranger missions and the Soviets' own lunar efforts, no one, not even the teams who built the probe, had much hope for this first mission. Gene Shoemaker, who led Surveyor

---

*   The Lunar orbiters mapped 99 per cent of the lunar surface in 1966–67. Really, this
         should have been done before Ranger flew, to ensure the missions were being sent
         to the best places possible. But, if there was ever a space programme to put the cart
         not just before the horse, but the driver and cargo too, it was Apollo.

1's camera team, gave a very pessimistic estimate of their chance of success – just 1 per cent, so you can imagine his shock when, 63.5 hours after launch, at 06:17 UST on 2 June 1966, Surveyor 1 bumped down onto the Moon's surface.

'My God,' Shoemaker exclaimed as the commentator announced the touchdown. 'It landed.'

The engineers had been gearing up to work out why the mission had failed, but now quickly reorganised their thoughts and prepared to analyse the spacecraft's data instead. The team immediately established that the lander was in good working order and communicating with Earth as it should be. Surveyor 1 had landed at a respectable speed of 3m/s, around walking pace, and rebounded just 6cm before coming to rest.

Once it was safely on the surface, the spacecraft got to work and its cameras gazed out at its new home. The first images were focused on itself, proving to those on the ground that the spacecraft had indeed landed in one piece. The vernier thrusters had cut out several metres above the surface, meaning the dust below the lander was undisturbed, but still the legs had only sunk in a few centimetres. As the bulky spacecraft had been scaled to mimic the pressure of the Apollo lander, it was a good sign that humans could land safely on the lunar surface.

Now it was time to get a real look around. The spacecraft's cameras swept across the landscape, revealing a gently undulating plain covered in craters of all sizes, with small rocks scattered between them. Several hills rolled across the horizon, although for the most part the surface seemed flat. Just what you would expect in a crater filled in by lava. Flat and free of large boulders, it was a far cry from the ragged landscape imagined by science-fiction artists of the day. It was, however, perfect for an Apollo landing. The mission had been a resounding success.

Surveyor 1 wasn't able to turn its neck to move its camera and instead relied on a movable mirror to look around. Although the camera was black and white, the team were able to use coloured filters to create a colour image, only to find the lander was the sole thing with any colour. The Moon was a world of grey rock and black shadow.

While the landing had proven that Apollo could set down on the surface, the human programme couldn't abandon its crew on the surface. They would have to take off again. The engineers fired up the vernier thrusters once more and the lander hopped up off the surface, not only proving the manoeuvre was possible, but the test gave the Apollo engineers enough information to work out how much oomph they'd need to get their future human astronauts back home to Earth.

As the first lunar day drew to a close on 14 June, Surveyor 1's camera watched as the Sun disappeared over the horizon. Just as the Sun dipped out of view, it managed to capture a fleeting glimpse of the corona – the halo of gas around the Sun that's usually only visible during a solar eclipse.

Without sunlight to charge its solar panels, Surveyor 1 went into hibernation through the two-week-long lunar night. When it woke back up again on 6 July, a week later than expected, the team cautiously began to check over their lunar explorer to see what had survived the -120°C temperatures of lunar night. While the core systems seemed to be up and running, the vernier engines had failed.

The next lunar night took out the cameras. The ailing spacecraft made it through a few more lunar nights, before it would eventually keel over on 7 January 1967 – a good innings for a mission almost everyone thought would fail.

Meanwhile, NASA was readying the next Surveyor mission to launch on 20 September 1966. It was bound for Sinus Medii and had been prepared early on the assumption that its predecessor would fail. It's perhaps poetic, then, that instead it was Surveyor 2 that failed. During its midcourse correction a thruster failed, sending the spacecraft spinning. After 29 hours trying to regain control, NASA lost contact with the spacecraft, and were forced to watch impotently as it mindlessly crashed into the mare it had been sent to explore.

Still, the programme carried on. On 20 April 1967 it was Surveyor 3's turn to get under way. Its destination was another lunar mare, the Ocean of Storms. From Earth to the Moon, everything went exactly to plan. When the time came to land, the spacecraft began its descent to the surface just as the NASA control team planned.

Everything was going smoothly. Then, as the spacecraft was just 10m from the surface, one of the radar beams Surveyor 3 used to judge the distance snagged on a particularly bright rock. As the confused guidance system struggled to correct itself, the lander came down at an angle. The on-board computer failed to realise the spacecraft had landed and fired the thrusters to correct the tilt. The lander bounced 15m upwards, about the height of a telephone pole, before crashing back down. The thrusters were still firing, launching Surveyor 3 a second time, this time reaching 11m. NASA, realising their spacecraft was about to bounce itself to death, managed to kill the engines remotely, but not before the lander bounded a third time, this time only reaching half a metre before finally coming to rest.

It was in a crater, tilting at a 12.5° angle – a bigger angle than Mission Control would have liked – but it was down. However, the less than ideal landing created some new problems. The turbulent landing had coated the mirrors in dust, blurring the camera's view, while the joints were caked in the stuff, hindering their movement. Several systems had short circuited. All these could be worked around but doing so would create a serious crunch in the schedule.

Despite its bruised appearance, Surveyor 3 sent home its first images within the hour. It lay halfway between a crater's centre and its rim. The location meant that the team could directly measure the depth of the hollow – around 20m. NASA located Surveyor 3 in images taken by their lunar orbiters, then estimated what they thought the crater should be by looking at the length of the shadow cast by the Sun in the orbital images – the longer the shadow, the deeper the crater. The calculation from the orbiter was 25m; 5m deeper than that measured *in situ* by Surveyor 3, but the exercise taught them to finesse how they estimated the depths of future craters. It is this kind of collaboration that makes landers and orbiters such a powerful combination when trying to understand another world.

Taking a closer look at the images, the science team were able to make out a set of scuff marks the lander had left in the lunar dust from its bumpy landing. This was a boon for the team responsible for looking at the properties of the lunar soil, as the bounce had cleared away the top layers of dust. The disturbed material was much darker than at Surveyor 1's landing site, suggesting that the upper layers were being cleaned or bleached to make them lighter.

For a more rigorous examination, however, Surveyor 3 was equipped with what would go on to be one of the staple elements of not just robotic landers but robotics in general – the mechanical arm. The history of the robot arm dates all the way back to 1495 and Leonardo da Vinci. This used joints, pulleys and cables in place of muscles and tendons to mimic the movements of a human arm, the basic principles that govern robotic arms to this day.

However, such arms are limited. Look at your hips, shoulders and ankles – they can move in all three degrees of freedom, thanks to their ball-and-socket joints. Recreating these complicated joints mechanically has proved extremely difficult. Instead, robot arms rely on simple hinges, such as those found at your knee or elbow. To mimic the wide range of movement of a ball-and-socket joint requires using several hinges pointing in different directions. More precision means more hinges, with more mechanisms to control them. As well as the logistical nightmare of operating half a dozen different moving parts at once, such arms quickly become too heavy to move.

Now, in the electronic age, the joints and pulleys could be replaced with electric motors, allowing the arms to become widely useful. The first industrial arm was installed at US car manufacturer General Motors in 1962. By the end of the 1960s, thousands of these arms were installed in production lines around the world, reaping the benefits of increased uniformity and reduced labour costs (although I doubt the workers who lost their jobs see that as a benefit).

At the same time, researchers realised such arms could be used in places where a human couldn't go. In the navy's case, that was under water; in NASA's, it was space. Surveyor 3 was their first attempt to land a spacecraft with a robot arm, specifically the Soil Mechanics Surface Sampler (SMSS). Rather than deal with complex hinges that would get incrusted with moondust, the Surveyor arm used a folding scissor design, similar to those you might see on cherry-picker lifts. It was flimsy compared to other designs and didn't give much control, but it was enough to get the job done.

On the end was a small scoop – nicknamed 'the Scratcher' – that could drag through the lunar soil, measuring how resistant it was to being dug out. Afterwards, the camera could photograph the trench to inspect the difference in appearance. Alternatively, the Scratcher would hold samples up to the camera for a closer look, or tip them out to see how they fell. The scoop could also serve as a makeshift geological hammer and move rocks around to examine them from different angles. With all of these methods combined, it was possible to tell a lot about how the lunar surface reacted to a good poking. The arm was meant to fly on a later mission, but after the success of Surveyor 1 there seemed no cause to wait.

The probe deployed its arm on 21 April and scraped its first trench a day later. At first, the going was fairly easy, but when the arm tried a second dig, things got much tougher. Under the initial fluffy layer of dust, it seemed the ground was much harder.

Surveyor 3 eventually dug out four trenches. These showed that the loose material covering the Moon was very fine and compressed down when the scoop was dropped onto it like a hammer. Importantly, the surface seemed sturdy enough to walk on.

Gene Shoemaker created the term 'regolith' to describe the rocky powder that covered the lunar surface, a word cribbed from the geological term for generic rock waste created by erosion. On the Moon, this erosion was caused by meteor impacts pulverising the larger rocks that used to make up the surface until the entire lunar exterior was covered in a layer of dust debris, a process known as gardening.

Perhaps one of Surveyor 3's more mesmerising sights was not on the Moon's surface but above it. Because of the lander's tilt, it was just able to get Earth in the frame of its camera. It was blurry, but clearly showed patches of cloud swirling over the globe – the first time our own world

The crew of Apollo 12 visit Surveyor 3. Astronauts Charles Conrad Jr (pictured) and Alan Bean dismantled the lander two years after landing and returned pieces of it to Earth. (NASA/NSSDC: nssdc.gsfc.nasa.gov/imgcat/html/object_page/a12_h_48_7133.html)

had been seen from the lunar surface. On 24 April, the spacecraft witnessed a lunar eclipse, although from that perspective it would be a terrestrial one, caused by the Earth moving across the Sun.

The lander shut down as lunar night drew in on 3 May, never to reawaken – although it wasn't the last time Earth ever heard from the lander. The site was deemed interesting enough for a return visit, and in 1969, Apollo 12 touched just under 200m away. Astronauts Charles 'Pete' Conrad Jr and Alan Bean visited the lander that had scouted their way. They took a few of the probe's parts – the camera, the scoop from the arm and several support struts – back to Earth to test how they'd fared during two years on the lunar surface.*

Surveyor 4 was intended to be a repeat of Surveyor 3, heading for the same region that Surveyor 2 had aimed for. Unfortunately, it suffered a similar fate to Surveyor 2. Two minutes before landing, the engine cut out. After Surveyor 2's failure, the spacecraft's design had been adapted to include diagnostic hardware to help prevent the problem from being repeated. Despite this, the cause of Surveyor 4's crash remained stubbornly unknown. Possibilities range from an exploding fuel tank to a broken transmitter – it's entirely possible that Surveyor 4 touched down safely and spent the rest of its lifespan waiting for Earth to call.

Despite the failure, NASA decided they'd done enough mechanical testing of the soil, and on Surveyor 5 replaced it with an instrument that could measure its chemical properties instead. It did this using an instrument called the Alpha-Scattering experiment. This contained a lump of radioactive material that emits alpha particles – basically the nucleus of a helium atom. The instrument fired these particles at the surface, then detected the pattern they bounced back in. By analysing the scatter pattern, ground scientists were able to work out what elements were in the soil, giving an insight into what the Moon was made up of (spoiler alert: it

---

*   At the time of writing, this is the only time that astronauts have paid a house call to one of the robotic explorers that paved their way (unless you count the fictional astronaut Mark Watney from Andy Weir's *The Martian*). With more human missions planned in the coming decade, this is a fact that could very soon be out of date.

isn't cheese). As the instrument wouldn't actually touch the lunar surface, there was no risk of it becoming contaminated by dust between readings.

During its journey to the Moon, Surveyor 5 had a few problems when pressure began to build up in the fuel regulators, meaning they had to burn some off to prevent the spacecraft exploding. There was still enough left to land, and on 11 September 1967, Surveyor 5 came down in the Sea of Tranquillity, 25km away from where Apollo 11 would land two years later.

The lander came down on a steep slope, causing it to slide before coming to rest at an angle – just like its older brother, Surveyor 3. As it slipped, its legs dragged through the regolith, creating a trench without using a mechanical arm.

During landing, the vernier landing engines cut out several metres above the surface, as usual. As well as ensuring the surface layer wasn't blown away, this kept the soil free of contamination from the engine's exhaust. Once the Alpha-Scattering experiment finished looking at this pristine top layer, Surveyor 5 fired its vernier engines, blowing away the loose dust to reveal the subsurface material. While this might have been contaminated by the exhaust, it would still give valuable insight into what the regolith was like when it was hidden away from the worst of the Sun's radiation.

The experiment found that the most abundant elements were the same as on Earth: oxygen, silicon and aluminium. It was also undoubtedly volcanic rock, solving the mystery of what created the lunar maria – these were no meteor craters.

Surveyor 5 had a rather more low-tech magnetic experiment, namely a magnet attached to one of its footpads. By looking at how the dust adhered to this, the team were able to tell how magnetic the soil was, giving an idea of its iron content. The pad had collected several grains of material, but much less than anticipated. The team had expected a lot of iron, brought in by the asteroids that pummelled the surface, but there turned out to be much less than predicted.

The lander fired its vernier engines one final time. The initial plan had been to fire them hard enough to relaunch the probe, but Surveyor 5's off-kilter angle meant it would have just fallen over, so the engines were throttled back. As they did, the lander was suddenly enveloped in a cloud of lunar dust. When the engines were on, the gas from the thrusters had diffused into the voids between dust grains. When they were turned off again, the gas rapidly escaped, creating a minor eruption. Fortunately, the spacecraft remained unharmed but dirty.

Surveyor 5 continued to monitor the Moon for four lunar days before permanently shutting down in December.

The overwhelming message from the first three successful missions was how universal the craters really were. Without labels, it was near impossible to tell photographs of the three apart. With the sixth outing, NASA wanted to land on something different, a plain, and once again headed back to the site of both Surveyor 2 and 4's demise – Sinus Medii.

Third time lucky, and Surveyor 6 touched down on 10 November 1967 without incident. Finally.

The plain Surveyor 6 found itself on was densely covered in craters. Using its own Alpha-Scattering experiment, the probe found that Sinus Medii's chemical make-up was almost identical to the Sea of Tranquillity's, suggesting that the geology of the lunar maria was pretty uniform. They were also strikingly similar to Earth in terms of minerals. That meant that the Moon didn't just form straight out of the solar nebula that birthed the planets – it must have gone through some form of complex geological processing just as the Earth did.

Despite the many craters, Surveyor 6 had managed to come down flat, so when the time came to fire up the vernier engines, the team weren't at risk of tipping the probe over. This time, they could really open up the thrusters.

Having learned from the previous mission, they throttled the engines up gradually to avoid raising a cloud of dust, and the 3m high Surveyor 6 gently lifted into the air. For the first time, NASA had (intentionally) relaunched from the surface of another world. The hop had only been 6 seconds long and carried the spacecraft only 2.5m from its initial landing sight, but the test proved that if the United States did ever land a human on the Moon, they stood a hope of bringing them back again.

Surveyor 6 managed to survive through the lunar night, making contact with Earth after sunrise, but fell quiet after only a few hours.

The purpose of the Surveyor missions had been to ascertain whether it would be possible to put a human on the surface of the Moon. The first four missions had shown it was possible to land on the surface and not sink. They'd also found that the older surfaces would be safer to land on – they had more craters, but that also meant the boulders had been gardened away. Finally, the tests with the thrusters had revealed the best way to relaunch once the moonwalking was done.

The Surveyor project had done its job of scouting the surface for Apollo. But there was still one probe left on the production line and so Surveyor 7 was handed over to the scientists.

As the mission didn't have to land on a 'safe' spot for a future Apollo landing, the mission could finally go somewhere interesting – they just had to decide where. The only thing the decision committee unanimously agreed on was that the lunar maria had been visited enough. Beyond that was a question of finding a site that was different enough to justify the trip.

One potential target could be trying to find the origin of several meteorites found on Earth, believed to have come from the Moon. They were far too acidic to have come from the volcanic maria but could instead come from the lunar highlands. Eyes turned to the 85km wide Tycho crater in the southern highlands. The crater was 4.5km deep, with a central peak that rose 2km over the crater floor. No one was sure if it was an impact crater or a giant volcanic caldera.

Landing in the terrain found at the rim of Tycho would be hazardous. Infrared measurements of the crater showed there were many hotspots, believed to be rocks sitting on the lunar surface soaking up the Sun's heat. The flight controllers decided the risk was worth the reward.

On 10 January 1968, Surveyor 7 successfully touched down, but the rough landscape took its toll on the spacecraft, seriously damaging one of the footpads. To maximise its science potential, the probe had been equipped with both an Alpha-Scattering and Soil Mechanics experiment. The Alpha-Scattering experiment could analyse the surface regolith, then the mechanical arm would clear away the dust without contaminating the soil with jet fumes.

The inclusion of both experiments would prove extremely fortunate. Shortly after landing, Surveyor 7 attempted to use the Alpha-Scattering experiment, but it wasn't deploying correctly. The spacecraft ended up using its scraping arm like a hammer, forcing the Alpha-Scattering experiment into position. It was an inelegant solution, but it worked.

The experiments found that in many ways the highlands were similar to the maria: although the landscape at Tycho was overall lighter, the surface was brighter than the subsurface and the soil was largely the same composition, although the mix of metals was a little different. It seemed the material that made up the highlands was slightly less dense, which made sense – if the Moon had once been covered in molten lava, the lighter material would rise up, creating highlands.

Surveyor 7 wasn't completely spared from Apollo, however. Several Apollo missions planned to leave behind reflector plates on the Moon.* These involved bouncing a laser off a device similar to the reflectors found on a car or a bike, allowing you to accurately measure the distance to the Moon. Astronomers used Surveyor 7 for a bit of target practice, shining a laser into its camera to confirm they'd hit it, which they did.

Controllers had to cancel a planned second hop due to overheating issues and the spacecraft entered hibernation on 23 January. It woke for a second lunar day, but its mission was largely over.

Surveyor 7 would be the United States' last robotic lander to the Moon. The Surveyor missions had been sent to scout out the lunar surface and lay the groundwork for Apollo. They had achieved that goal.

On 20 July 1969, the Eagle landed in the Sea of Tranquillity and Neil Armstrong took one giant leap for mankind. For the United States, the lunar surface was now the domain of human explorers, not robotic ones.

---

*   Apollo 11, 14 and 15 left behind these reflectors, as did both the Soviet lunar rovers that we'll cover in Chapter 6. People still use them to track how far away the Moon is, as well as proving to moon-landing deniers that at least something human made is on the lunar surface.

5

TO THE MOON AND BACK

The Surveyor missions threw the rudimentary nature of the Soviet lunar efforts into sharp contrast. Although the Soviets continued their work at the Moon, there was no denying that the United States' probes were the better hardware.

After Luna 9 made the first landing, the Soviets focused on orbiter missions before returning to the surface with Luna 13 on 24 December 1966. The spacecraft was based on the same Ye-6 design as Luna 9, upgraded with two arms that unfurled like extendable ladders, one of which carried an instrument to measure the density of the lunar soil called a penetrometer. Compared to the instruments on Surveyor, Luna 13 was woefully simplistic.

In terms of a human lunar landing, the Soviets were falling even further behind. With no centralised plan and no Korolev at the helm, the Soviets had ended up with at least two programmes developing human space-flight. The competition between the two teams was holding them both back from getting anywhere, and all their squabbling seemed to do was waste time and resources. Meanwhile, several core technologies – such as the Zond module that could carry two cosmonauts and the N1 rocket needed to launch them – had met with failure after failure.

The Soviets had been leading the Space Race so far, but it was clear that the United States was about to surge into the lead. The decision to publicly deny the existence of a lunar programme had proved wise. There was no way that the Soviets would be able to beat the United States to land a human on the Moon. But if they couldn't win the game, there was a chance they could change the rules. It might be possible for the Soviets to bring home some of their own moonrocks before Apollo 11 could and do so without sending a human to the surface.

When the United States announced the dates of the Apollo 11 landing, the Soviets decided a probe of their own should run alongside it. Luna 15 would launch on 13 July 1969, just three days before Apollo 11 left Earth. It would land on the Moon an hour before the Apollo 11 crew. As Neil Armstrong took his 'one small step for man', the Soviet lander would already be hard at work, collecting a lunar sample. Three hours after the Eagle took flight again, Luna 15 would also set out for home, carrying its own sample of moon rock back to Earth. It would arrive on Earth just 2 hours after the Apollo 11 crew.

It was an audacious plan. If it succeeded, the Soviets could use it as a catalyst to reframe the entire Space Race: the wasteful capitalist society of the United States spent billions of dollars and risked human lives to do what the Soviets could do cheaply and safely with a robot.

The Lavochkin bureau had already been working on a sample-return spacecraft, the Ye8-5. Even though the design – a pyramid of bulbous white chambers and instruments balanced on four slender, silver legs – was relatively spacious, there was still a limit to what it could carry and, therefore, what it could learn on the surface. Instead, they wanted to bring a piece of that rock back to Earth, where it could be experimented on in the best laboratories in the world.

When the Soviets attempted the first test flight of a Ye8-5 in June, they failed to even get out of orbit. The problem was the Proton launch rocket. Out of seventeen launches, the Proton had failed nine times. There wasn't time to design an alternative, and so as the world's excitement built for the launch of Apollo 11, the Soviet team were crossing their fingers, hoping their own efforts wouldn't be scuppered by a broken rocket. Their prayers were answered on 13 July, when the spacecraft set off on its way to the Moon.

With usual Soviet reticence, the government didn't say exactly what the mission was doing. Although many in the West guessed its true purpose, theories ran wild: it would jam Apollo 11's communications; it was a rescue mission if they failed; it was a spy satellite to ensure Apollo 11 did what NASA said it did.*

----

*   This would have cleared up a lot of conspiracy theories if this had been the case.

As usual, Jodrell Bank watched the spacecraft's progress and listened in on 17 July as it approached the Moon. It seemed all was not well – its orbit was far too elliptical. Although the Soviet press proclaimed everything was fine, in reality Luna 15 had failed to reach the correct orbit and Ground Control were struggling to wrestle it back into something resembling the correct position.

On 19 July, Apollo 11 arrived in orbit. With the world's attention now focused on the humans about to descend to the lunar surface, Luna 15 itself prepared to land. Unfortunately, the Soviet control room was a scene of chaos. The radar that would guide the spacecraft down to the surface found that the terrain was much rougher than expected. The mission was delayed by 18 hours to get a clearer picture. It meant Apollo 11 would beat them, both to the surface and back to Earth, but if it would save the mission it was worth it.

Luna 15 met its eventual fate just as Neil Armstrong and Buzz Aldrin were preparing to return from the lunar surface. The Soviet probe was meant to fire its thrusters for 6 minutes, but the spacecraft crashed into the Moon after just 4, when it should still have been 3km away.

According to Jodrell Bank, the spacecraft crashed into the appropriately named Sea of Crises at a speed of 480km/h. The Soviets tried, again, to pass off the mission as an intentional impact but few were fooled.

Had it succeeded Luna 15 could have completely changed the way we view the Apollo missions. Apollo was grand and inspiring, there's no doubt, but it was also horrendously expensive and dangerous. Even without Luna 15 supplying an alternative method of returning lunar samples, the public perception of Apollo was beginning to shift. Although no longer alive to see it, JFK had had his big moment – but now that moment had passed.

The 'Moon fever' of the US public faded, and the western media grew disenthralled with human lunar travel. The Space Race had reached the finish line. Without that sense of competition and glamour, the enormous cost of Apollo was hard to justify.

Interest briefly returned when Apollo 13 suffered a catastrophic failure en route to the Moon. The rescue showcased NASA's talent in 'bringing their boys home safely', but the near tragedy reminded people how dangerous human lunar exploration was. These were real lives the government was risking at a cost of several billion dollars. When Apollo 17 flew to the Moon in December 1972, the decision had been made – NASA's time on the lunar surface was over.

The Soviets, by contrast, weren't as quick to give up on our celestial neighbour. Luna 15 might have failed, but the idea of robotically returning samples with a lander Ye 8-5 was still a good one. With the human Space Race lost, the Soviets could take their time with the robotic one.

After Luna 15, a second round of sample-return missions was ordered and on 12 September 1970, Luna 16 began to make its way to the Moon's surface. It was largely unnoticed in the West as the launch coincided with Black September, a conflict between Jordan and Palestine that resulted in several civilian aircraft being taken hostage. Although understandable, it's a shame that Luna 16 has been forgotten by history.

Luna 16 slid into orbit perfectly. As it began to scan the landing site, mission controllers were relieved to see that it was wonderfully flat. Although the Sun was still below the horizon, the spacecraft made its way down to the dark surface on 20 September, making a textbook landing. The cameras switched onto reveal the dawn-lit lunar landscape.

Eager to get a sample of moonrock, Mission Control deployed the spacecraft's drill. Luna 16 could bore down through 90cm of rock but stopped when it struck something hard. Rather than risking the drill burning out or, worse, getting stuck, Luna 16's controllers ordered the spacecraft to stop drilling. They just had to pray the drill had done enough to at least get something. Luna 16 bundled up its precious cargo, however much of that it might be, and prepared to head back home.

After just 26 hours on the surface, explosive bolts detonated, separating the return capsule from the lander, and a jet of flame propelled Luna 16 back home. Several days later, the capsule struck the Earth's atmosphere. Enduring 10,000°C temperatures and 350G forces, the capsule crashed down in the region that is now Kazakhstan. With its ordeal survived, the probe began screaming out a radio signal, calling the recovery team to come and find it.

When Soviet scientists finally cracked open the capsule, they found a small sample, weighing just 101g – tiny compared to the 20kg of rocks brought home by each of the Apollo missions. But it was theirs. It was a triumph for the mission.

'Automatic devices are now designed to conduct the main part of the exploration of outer space, the Moon and other celestial bodies of the

Solar System,' said leading Soviet space scientist, Boris N. Petrov in an interview widely disseminated by Tass, the official Soviet press office. 'The flight of an unmanned craft compares in cost to a manned one by a factor of one to 20 or 50.'

The Soviets successfully returned to the Moon's surface on 14 February 1972, when Luna 20 landed on the lunar uplands. However, once on the surface, things didn't go as well. The drill had trouble boring into the rock and had to stop three times to prevent overheating. After nearly 3 hours, the drilling was halted. It being uncertain exactly how successful the mission had been, the spacecraft was instructed to seal the return capsule and make its way home.

It arrived back on Earth on 25 February into the middle of a blizzard. The recovery helicopters were forced to wait out the storm as the probe was blown towards the Karakingir River, in a remote area of west Kazakhstan. After making it safely back from the Moon, would these samples end up buried at the bottom of the river?

When the winds abated the next day, the probe was found, battered and burned but whole, on an island in the middle of the river. The sample was small, less than 50g. But moondust was moondust, and the Soviets treasured it just the same.

Luna 23 made its way to the Moon in October 1974 but was severely damaged on landing and could not collect its sample. The next mission failed to even leave Earth orbit. After everything, were the Soviet efforts to the Moon about to end on a sour note?

The final mission, Luna 24, broke the streak of bad luck. When it arrived at the Moon on 18 August 1976, it came down in the exact spot Luna 15 had meant to land. It returned to Earth less than a day later, crashing into Siberia. This time, the capsule was easily recovered, and the team found an impressive 170g of lunar regolith inside. A deceptively straightforward end to a campaign that had been far from simple.

Years later, as a message of peace, the Soviets exchanged 3g of Luna 16 moonrock for 3g of Apollo moonrock. A similar exchange was made for Luna 20 samples. Later still, when the Soviet Union was facing financial difficulty, 2mg of the soil sold at auction for $442,500, before later being sold on for $855,000. Once again, it would appear the value of the lunar missions was far more than just scientific exploration.

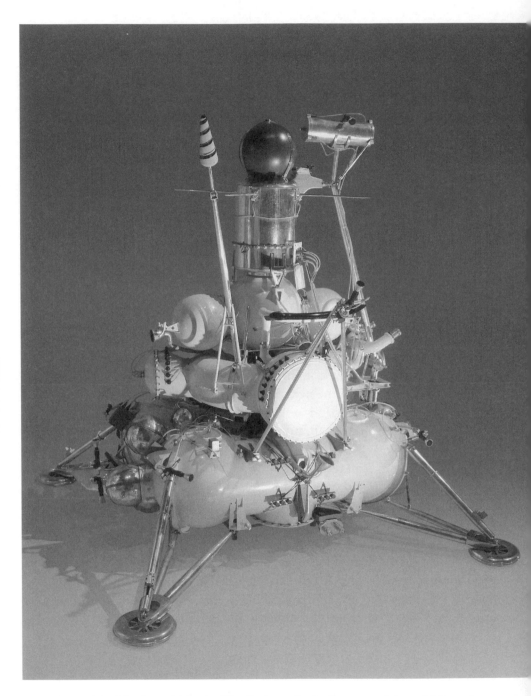

A model of the Luna 16 lander. The Soviets robotically returned material from the Moon three times using spacecraft based on the same design as Luna 16. (Elizabeth Pearson/Science Museum London)

Alas, these auctions were the only times the West remembered Luna 16, 20 and 24. Despite being the most successful robotic sample-return missions ever mounted, they have largely been forgotten. The Soviets had mounted Luna 15 in an attempt to rewrite the rules of the Space Race in favour of robotic spacecraft. While the rest of the mission might have been a success, that goal undoubtedly failed.

6

MOON ROVERS

With lunar rock of their own to examine, the Soviet scientists could now begin analysing the regolith with high-level instruments that would never fit on a planetary lander. Yet, even the best labs on Earth cannot replicate conditions on the Moon. Moisture in the air – in fact, the presence of air full stop – makes lunar dirt behave differently, as does Earth's higher gravity. If you really want to fully understand how regolith acts, then you have to look at it *in situ*, on the Moon.

If the Soviets were going to continue exploring the lunar surface, then they would need to upgrade their hardware. They could simply attempt to mimic the Surveyor probes, but this would seem to be a step backwards and they would still suffer from one of the probe's biggest drawbacks – they were stationary, bound to only explore where they happened to come down. The Moon is a big place, filled with interesting sites and locations and the Soviets wanted to explore them. Just as they had sought to bring home samples with a robotic explorer, they now sought to roam around the surface with one too.

On 17 November 1970, word came in that a Soviet mission, Luna 17, had set down in the Ocean of Storms. Just over a year after the United States had sent its first moonwalkers, the Soviets sent their own, albeit a robotic one – a rover named Lunokhod, literally translating to moonwalker.

The rover was a strange-looking thing, somehow both bulky and delicate. Standing 1.5m tall and resembling a metal bathtub on wheels (albeit one daubed with the sickle and hammer), it had been built to survive the harsh environment of the lunar surface. During the night, its top was covered over with a solid lid that, come sunrise, opened up to reveal a solar panel and a large flat radiator to expend the sun's heat. However, its eight wheels – four on each side – seemed to have been made from chicken wire, with spokes so fine as to be invisible.

Lunokhod was a robotic scientist, with several instruments ready to start examining the surface. A camera looked around the surface to navigate, while an X-ray spectrometer – a device that used X-rays to measure the elements within the lunar rock – and a penetrometer investigated the physical properties of the lunar regolith.

The Lunokhod programme had been in the works since the 1950s. Initially, the rovers were to serve as precursors to crewed missions, with plans to set them down near potential landing sites as an escape route should everything go wrong.* But, as it became apparent the Soviets would never send a human moonwalker, they instead decided to send a robotic one.

Unlike the Luna landers, Lunokhod's mission was meant to last not one lunar day, but many. That meant it had to survive the lunar night. To keep the spacecraft running, the rover contained a lump of radioactive polonium, not to act as a power source but as a 'hot water bottle', keeping the rover toasty through the night.

The first attempt to land a Lunokhod was made in 23 February 1969 – another attempt to rewrite the narrative of lunar exploration by beating human moonwalkers with a robotic one. But it wasn't meant to be. The ever-problematic Proton rocket exploded after less than a minute. The rover's remains rained down on Earth just 15km from the launch site.**

---

*   Many in the West believe the Soviets were cavalier in their regard to human life within the space programme. In reality, they were far more risk averse than their US counterparts. While Apollo had been rushed through, doing only one test flight before moving onto the next iteration, the Soviets had insisted that there would have to be multiple successful tests before any human lives were risked. Sending an escape route to the Moon ahead of time was just one of many ideas the Soviet lunar programme had to keep their cosmonauts safe.

**   Knowing that the spacecraft debris would include a lump of nuclear material, the military set out into the Russian wilderness to track it down. Legend has it that the radio thermal generator (RTG) turned up in a patrol hut where local troops were using it as a heater. It was quickly taken somewhere a little safer. Other parts of the rover survived remarkably undamaged, including a portable tape recorder. It was supposed to play the Soviet national anthem as the rover rolled down onto the lunar surface. It was still playing the tune when it was found on the Steppes, albeit rather forlornly.

A year and half later, Luna 17 made a second attempt and met with considerably more success. It came down on the opposite side of the Moon to Luna 16, in a region called the Ocean of Storms, in the Sea of Rains basin. With advanced computing still to be invented, Lunokhod 1 was operated entirely from Earth. Two teams of five drivers controlled the rover's every move, alternating shifts through the lunar day and catching up on sleep through the Moon's night. It was a demanding job, requiring intense focus with the knowledge that even one wrong command could destroy their extremely expensive charge. The drivers were carefully chosen to make sure they could stand up to the stress.

To make matters worse, the drivers' first job was one of the most dangerous of the entire mission – driving off the landing platform. The rover sent back images of the ramps that would lead the rover to the surface. It looked as if the way ahead was clear. And so, on 17 November 1970, the driving team began their long mission, and coaxed Lunokhod 1 down onto the surface.

The view of the Luna 17 lander from Lunokhod 1. The rover's drivers had to use these images to navigate the lunar terrain. This was made difficult by the lack of contrast and low angle to the ground. (Don P. Mitchell/ Roskosmos/ Russian Academy of Sciences)

However, rather than being given the peace and quiet they needed at this tense time the drivers were subjected to a crowded control room. The landing had created a new flurry of interest in lunar travel, both in the Soviet Union and in the West. Wanting to make the most of the propaganda opportunity, the control floor had been opened up to any journalist who cared to visit. Instead of being allowed to learn how to drive on another world in peace, the drivers were constantly subjected to the shouts, 'Watch out for that rock', or calls of, 'You're going to crash!'. It was only when the drivers' heartrates began to climb over 140bpm that the director of Lavochkin, Georgi Babakin, decided enough was enough and kicked everyone out.

The backseat driving was particularly galling as operating Lunokhod 1 proved far from easy. The navigation cameras were low to the ground and this was not the best perspective to navigate from. As the images were black and white, the drivers could only make out craters and rocks by their shadows.

At the best of times, it was difficult to make anything out. At high noon, when the Sun was overhead, there were so few shadows that driving became impossible and the rover had to stop for several Earth days. To make matters worse, these images were only sent back every 20 seconds. While the rover was driving very slowly, meaning that there weren't huge jumps between images, the drivers couldn't react to events as they happened and had to remember their planned route. Then, early on, the brakes failed so that they were permanently being applied, forcing the drivers to work against friction for almost the whole mission.

The going was slow, and when it shut its lid to conserve heat through the night, it had barely moved away from the lander. But with every day, the rover team began to get the hang of driving on the lunar surface. Navigation continued to prove difficult, needing the drivers to not just avoid obstacles but to keep the solar panels lit and the antenna pointing at Earth.

All the same, the miles began to rack up. By the end of December, Lunokhod 1 had lumbered 1.4km away from the landing stage, taking images of the stunning surroundings. The further the rover went, the more obsessed people became with the mileage count, especially the journalists at *Pravda*. The obsession fed back into the control room, with the driving team being pushed to travel further each day.

Meanwhile, the science teams were being pushed to the side. The drivers wanted to stay away from any dangerous features that could harm the rover, such as rocks and craters – the exact kind of thing the scientists wanted to investigate. Eventually, principal lunar geologist Alexander Basilevsky begged Babakin to let them stop and look at some of the interesting rocks the rover passed. Babakin is reported to have shot back it was called Lunokhod, not 'Lunostop'.

What had been the point of sending this incredibly advanced scientific machine if you wouldn't let it do science? On one occasion, Basilevsky requested permission to take a stereoscopic lunar panorama, but was told by the colonel in charge that it was unnecessary – it was just another boring lunar scene. In frustration, Basilevsky set up one of Lunokhod's previous stereo panoramas in a special viewer to make the lunar surface 3D. Confronted with the stark beauty of the scene, the colonel consented. Slowly, the scientific team were making their case heard and, in time, found a balance with the rover operators to keep the rover safe while also getting the chance to use its instruments to collect data on the lunar surface.

One such instrument was the X-ray spectrometer. This investigated the chemical composition of the soil much like Surveyor's had but used X-rays instead of alpha particles, allowing it to detect a greater range of elements. It found the regolith rich in aluminium, calcium, silicon, iron, magnesium and titanium – this was typical of a mare area.

But the real advantage of Lunokhod was being able to look at how the lunar regolith physically acted in its natural environment, focusing on tests that couldn't be run in the labs back on Earth. At regular intervals, Lunokhod used a mechanical rod, named PrOP, to hammer the soil, just as the mechanical arm had done on Surveyor, and this revealed that the regolith in the Ocean of Storms was weaker than elsewhere.

As 1971 dawned, Lunokhod doubled back towards its lander before heading north. The rover drove along crater rims and even ventured inside them, eventually coming to a field of craters strewn with rocks. Although the drivers wanted to avoid the hazard, the scientists campaigned to visit them. The scientists won, but it turned out the engineers' fears were well founded. On 13 April, the rover became stuck in a pile-up of dust. It managed to pull free but used up so much of its power supply it had to spend the rest of the month recharging.

By the end of May, the rover was beginning to lose power with each charging. Given that it had only been intended to last three months, and had already survived double that amount of time, there were few complaints as the team prepared for its slow demise. The rover was still running on 30 July when the crew of Apollo 15 flew overhead, bringing their own rover to the surface, admittedly one driven directly by human hands (specifically those of moonwalkers David Scott and James Irwin).

The end came on 4 October when Lunokhod's heart gave out. There was a sudden pressure drop in the sealed interior that held the most sensitive instruments at Earth room temperature and pressure. First, Lunokhod lost the use of its wheel, then its sight, and finally fell silent.[*]

It had lasted ten months, travelled 10.54km, taken twenty-five X-ray readings and broadcast over 20,000 pictures. After months of hard work, the drivers left the control room for the last time.

The Soviets would soon be back, as the second Lunokhod was already being built. On 8 January 1973, a month after the last Apollo mission, Luna 21 sent the last moonwalker – Lunokhod 2. It touched down in the relatively flat LeMonnier Bay on 16 January.

Having learned from Lunokhod 1, the cameras were positioned higher on the rover and could transmit images back every 3.2 seconds, making navigating much easier. The science suite had been given an upgrade too, now featuring a magnometer to hunt for signs of a magnetic field, a laser that would be able to accurately measure the distance to Earth and even an experiment measuring the level of ultraviolet light coming from our galaxy, investigating the potential for using the Moon as a base to set up an astronomical telescope.

As Lunokhod 2 waved goodbye to its landing stage, it set out towards a mountain range 7km to the south – a region that had conveniently been photographed by the Apollo 17 astronauts a month earlier. By the end of its first lunar day, the rover was already 1km of the way there. Later on, it was managing that far in a single Earth day and at the end of the second lunar day it had already beaten the record set by Lunokhod 1's whole mission.[**] In fact, it was proceeding at such a pace that when it started moving on the second day, the rover refused to stop. The drivers sent the halt

[*] The final location of Lunokhod-1 remained unknown until 2010, when astronomers succeeded in bouncing a laser off its retroreflector.

[**] The drivers were so into the swing of things they sometimes lost track of time. One evening, the rover's signal suddenly cut out. Fearing some catastrophic failure, the team began to try get the signal back. It was only when someone popped outside and noticed that the Moon had set, passing out of radio range, that they realised what had happened. They'd all been too 'in the zone' to notice the time!

command three times before Lunokhod 2 finally listened and stopped itself plummeting into a crater.

When the rover arrived at the mountains in April, it began to explore a rille, a long, narrow channel that can be several kilometres wide and hundreds of kilometres long. It was a dangerous journey, as the area was strewn with boulders – too dangerous. When the team attempted to wake the rover at the start of the June lunar day, it quickly and quietly died.

The culprit was eventually found to be some dirt that had fallen on the lid the previous lunar day, presumably when the rover had brushed against a rocky wall. When the lid closed for the night, this light layer of dust fell on the radiator. As daytime rolled around again, the lid was lifted, but the dirt remained, insulating the radiator. With no way to lose heat, the temperature rose quickly, killing Lunokhod 2.

Despite its shorter life, Lunokhod 2 outperformed its predecessor. It travelled 37km, took 80,000 images and picked up traces of a weak magnetic field.

Perhaps its most interesting findings were not related to the Moon's surface but to its sky. Hopes of a lunar-based telescope looked unlikely to see fruition, because while the lunar nights were brilliantly dark, during the lunar days dust blocked the view, making it hard to observe. It would also be quite a bright location as the sunlight reflecting off Earth would mean a lunar stroll was fifteen times brighter than a moonlit walk on Earth.*

A third lunar rover was built but never flew. Lunokhod 3 is now a museum piece, a reminder of an impressive, but often overlooked Soviet lunar programme. Between Lunokhod and the sample-return missions, the Soviets had proved that robotic missions could explore the lunar surface and bring parts of it home.

Not only were these missions cheaper than human ones, they were safer, too. This last point was hammered home during Lunokhod 1. Early

---

*   Lunokhod 2's story doesn't end in 1973. Twenty years later, when Russia was desperate for cash after the fall of the Soviet Union, they sold the rover for $68,500 to space tourist and video game developer Richard Garriott, making him the only (at time of writing) owner of an object on an alien world. As Garriott is a private individual and not a government, he is not technically subject to international law governing who can lay claim to territory in space. In a 2001 interview, he jokingly claimed dominion over the Moon in the name of his gaming alter ego, Lord British.

in 1971, the rover was struck by a solar flare – an outburst of particles from the Sun. The radiation from the flare had been strong enough that it would have injured any human who'd been in the hardy rover's place.

With no Lunokhod 3 to travel to the Moon, the lunar Space Race ended with Luna 24. By the end, both sides had learned a valuable lesson. The thrill of competition, although providing both sides with the drive to forge on, had led to people rushing when they need not have done, leading to costly mistakes.

Yes, the United States had put a human on the Moon, but at an immense financial cost and with little idea of the way forward now. Meanwhile, the Soviet Union, whose real ambitions lay in exploring the planets, had let themselves be dragged into a race that they not only couldn't win, but they'd never even wanted to run in the first place. It was rapidly becoming apparent that the Space Race had been a competition where everyone lost.

Direct competition was a game where no one won; instead of being competitive, space missions should complement each other. It was time for the Soviets to go back to what they really wanted to do, and what they had been doing in the background throughout the lunar race. They started with Venus.

PART 2
VENUS

7

THE JOURNEY TO VENUS

The Soviet populace have long been enamoured with our two nearest planetary neighbours, Mars and Venus. It was an obsession that began with the work of one man, now known as the father of Soviet rocketry. His name was Konstantin Tsiolkovsky.

Born in 1957 to a poor Polish immigrant family, Tsiolkovsky was left deaf at the age of 10 following a bout of scarlet fever. Unable to attend school, he became a voracious reader, teaching himself mathematics, physics and chemistry. His biggest inspiration was the works of Jules Verne. He was stirred by the stories of adventurers firing themselves out of a cannon towards the Moon and began working out how you could feasibly venture out into space, publishing many of his calculations in Soviet scientific and aerospace journals.*

In 1903, now working as a provincial school teacher, Tsiolkovsky published his most influential work, *Exploration of Outer Space by Means of Rocket Devices.* The paper described a rocket powered by liquid oxygen and hydrogen, and the path it could take to reach our neighbouring planets. The ideas were decades ahead of the technology needed to realise them but outlined what would become the basis of modern rocketry.

Due to Tsiolkovsky's relative obscurity, his work wasn't widely translated until after his death in 1935, by which point others had independently come to similar realisations, such as Hermann Oberth in

---

* He even went so far as to build a centrifuge in his back garden to test his theories on how changing gravity would affect a living creature, in this case some of the local chickens he'd rounded up.

Germany and Robert Goddard in America. Within the Soviet Union, however, Tsiolkovsky's work seeded the idea of travelling to other worlds in people's minds.

As the century wore on the idea of spaceflight, particularly to Venus, continued to grow in the public consciousness. Although both Mars and Venus are about the same distance away, Venus is similar in size to Earth. With a thick cloud layer obscuring the surface, science fiction writers and scientists alike had long imagined a fantastical world hiding from view.

On Earth, cloud cover is caused by a damp climate, and so many assumed that Venus was a wet and swampy world. One astronomer, Gavriil Adrianovich Tikhov, fuelled these ideas in 1947 when he set up the world's first astrobotany department at Alma Ata Observatory in Kazakhstan, to study the plants he felt certain were growing on both Mars and Venus. Tikhov spent years travelling the world to find life in the most extreme environments. If life could survive in the frozen Arctic or a boiling geyser, then why not on Mars or Venus?

As the Space Age wore on, Tikhov's work began to shine a light on the second planet from the Sun. He carefully observed Venus to measure its climate, showing it was between 60–76°C. This was hot compared to Earth but still cool enough to allow for liquid oceans. However, in the 1960s that idea began to falter. Radio astronomy suggested that Venus could be much hotter, but there was no consensus. Estimates ranged from Earth temperature all the way up to 300°C. If the top end was true, the world would be baked dry.

There was more to studying the planet than searching for an alien civilisation beneath the clouds, however. Venus is, in many ways, a similar planet to Earth. It's closer to the Sun, but not by a lot, and is almost the same size, mass and density. Venus offers the chance to study a planet that is almost identical to Earth – a great place to test out theories about how planets form.

Our current understanding of planet formation is that they grew from a disc of dust – the protoplanetary disc – that surrounded the infant Sun. Close in to the Sun, temperatures in the disc got hot enough to melt, or even vaporise, elements like iron and silicon. This means they could easily interact with each other to form complex minerals like iron oxide and silica. As the solar system began to cool down, these minerals clumped together to form droplets, which then came together to form rocks, growing into bigger rocks and so on, until eventually you ended up with something the size of a planet. As time went on, there was some shuffling around, with planets moving closer and further away from the Sun as they jostled themselves into a stable position, eventually arriving at the solar system we see today.*

As different metals have different melting points, and because it gets cooler the further away from the Sun you are, the mix of various minerals on offer in this protoplanetary disc change depending on how close to the Sun things are. This means the basic chemical make-up of each planet could be different, depending on where in the protoplanetary disc it formed. It's only by comparing all the planets in our solar system that we can really understand how any of them came to be. The differences between Venus, Earth and Mars tell us about the conditions that made them. In learning about these other worlds, we begin to unpick the mystery of our own.

However, there is only so much you can learn about your neighbours by spying on them through the window, especially if they're hiding behind their curtains in the way Venus hides beneath its clouds. The only real way to get to know them is to knock on the door and pay a visit.

After Sputnik in 1958, when Korolev was formulating plans to visit the Moon, he made moves to begin missions to Mars and Venus, starting as early as 1959. The ultimate goal was to send human explorers to both planets, but robotic emissaries would scout the way first, just as they had on the Moon. Initially, these would be flyby missions, with the intention of sending a lander later.

There wasn't much time to plan a mission. Unlike the Moon, which orbits Earth, Venus makes its own way around the Sun. In fact, it only comes close to Earth once every nineteen and a bit months. There's not much point wasting time and fuel flying to a planet when it's on the other side of the solar system, not to mention the problems in trying to keep in radio contact with a spacecraft when it's behind the Sun. Instead, Venus missions would have just a few weeks when the planets were lined up where they could launch. The next such launch window was in June 1959. They needed to get cracking if they wanted to make it.

Despite their best efforts, development didn't go well. As the launch date grew close it became apparent that any spacecraft the Soviets produced wouldn't be up to standard. While a high failure rate had been acceptable for the lunar flights, which could be launched every month, messing up a Venus mission meant waiting a year and a half for the next chance. Rather

---

*   In the outer solar system things were a bit colder and worked slightly differently, but
      we'll come to that in Part 4.

than rush through a spacecraft that would almost definitely fail, the team delayed the launch until the next window in 1961, giving them time to build one that could actually work.

With missions planned to Mars around the same time, the Soviets designed a universal interplanetary spacecraft dubbed 1MV (for Mars Venus – the Soviets have always been very practical when it comes to naming things). Like the original Ranger plans, the 1MV would be multipurpose, capable of going to either of the planets as a flyby mission and could be adapted to carry a lander as well.

The 1961 round of Venus probes were intended to carry landers. However, touching down on Venus would be a more complicated affair than on the Moon. Above all, it has an atmosphere. When the spacecraft first arrived at the planet it would be travelling tens of thousands of miles per hour. At those speeds, even the diffuse atmosphere at high altitudes would generate a huge amount of frictional heating when the probe slammed into it.

Fortunately, the spacecraft could also use the air resistance to slow down, provided it had a heat shield to keep the hot air away from the spacecraft itself. Once the lander slowed down enough and dropped into the thicker atmosphere lower down, it would be able to use parachutes to carry it the rest of the way to the surface without the need for retrorockets. As Venus's gravitational pull is about the same as Earth's, it would fall much faster than on the Moon. This could be compensated for with a bigger parachute – but a bigger parachute meant a longer fall time. As the lander was running on batteries, there was a chance it could run out of power before ever reaching the surface. It was a delicate balancing act. If they did luck out and managed to survive the descent, the lander was designed to cope with both a hard landing on a solid surface and a splash down in one of the many oceans believed to cover the surface.

Even with the extra eighteen months, Korolev was overstretched and the design came out rather slapdash. Combined with the fact that the landing team only had a fuzzy idea of what the atmosphere was like, and so were unlikely to get the balance of parachute size correct, no one expected the mission to get far. In October 1960, two Mars probes failed to even make it out of orbit, so the decision was made to scale the Venus project back to a flyby.

The first Venus probe launched on 4 February 1961, only making it as far as parking orbit around Earth. As usual, the Soviets hid the failure by claiming they were testing a new variety of heavy satellite. The West, however, suspected it was a first attempt at crewed spacecraft that had gone tragically wrong. The spacecraft crashed back to Earth shortly afterwards.*

Fortunately, the Soviets already had a second probe ready. The philosophy was that while one probe might fail, the chance of both not making it was much smaller. As the lion's share of the cost of spaceflight is research and development, building an extra probe is a cost-efficient way of dramatically increasing a project's chance of success.

The back-up mission launched on 12 February, and things seemed to go well as it made it out of Earth orbit. For thirteen days it sped on towards Venus, before the spacecraft stopped reporting in. It seemed the early design wasn't good enough to reach the planet.

The team at Jodrell Bank tracked the spacecraft as it journeyed towards Venus, measuring that it passed around 100,000km away from its target. The spacecraft had been off course and silent, but it had flown past Venus, earning it the title Venera 1.** Although it was itself a failure, it was the first mission of what would go on to be one of the early Space Age's most successful campaigns.

With another nineteen months until the following window, the team began to work on the next iteration of spacecraft. Again, the mission was kept generic, but this time it came in three varieties. The 3MV1*** would be a planetary impactor, the 3MV2 for flybys and the 3MV Zond**** would be a test bed, used in the long gaps between launch windows.

The multi-use nature of the spacecraft was achieved by splitting it into several key components that could be interchanged and developed as

---

*   Just like the lunar landers, the probe contained a ball of pennants to be scattered across Venus's surface. These were found two years later and handed over to the authorities. They were sold in 1996 to raise money for the struggling Russian science programme.
**   Initially, the spacecraft was called the Automatic Interplanetary Station, a generic name given to several interplanetary craft. It was given the Venera designation later in recognition of its achievements. To keep things from getting confusing, I will use the Venera designation throughout.
***   The eagle-eyed among you might have noticed the lack of a 2MV series. We'll cover that iteration in Chapter 10. Just be assured it was not a great machine and was almost instantly replaced with the 3MV.
****   The Zond designation caused no end of confusion. The West assumed it was a designation meant to cover up failed deep-space missions, as the designation Kosmos had done for those in Earth orbit. To make matters worse, the moniker 'Zond' had also been used for several crewed flights. It's no wonder there is still such confusion over exactly what the Soviet space programme did in these years.

needed. On one end of the spacecraft were the propulsion systems. In the middle was the central 'stack', known as the orbital compartment, with all the parts the spacecraft needed to function – the electronics, the communications and the power source attached to two solar panels with a wingspan of 4m. Below this stack was the planetary compartment, which would change depending on exactly which iteration was being used.

The 3MV1 had a spherical lander around 1m wide. Usually painted white, these resembled giant ping-pong balls with several antenna and spring-shaped instruments jutting out of the top. The 3MV2 would replace the lander with remote sensing instruments, such as radar. It would ultimately turn out to be a solid design and was used for every Venus mission between 1964 and 1972.

The first launch of the 3MV series came in the spring of 1964 with three missions aiming to travel to Venus. The first (a 3MV2 flyby) blew up during launch and the second (a 3MV1 lander) reached orbit but failed to get any further. It turned out that a gyroscope – a device to help keep the spacecraft orientated – had been set with too narrow a margin of error. Irritatingly, it took the engineers just 15 minutes to fix the problem, which they did for the third attempt (another 3MV1 lander).

Finally, this managed to get under way towards Venus on 2 April 1964. But shortly into the flight a welding seam cracked, exposing the spacecraft's electronic guts to vacuum. The internal workings of spacecraft are designed to work at room temperature and pressure, then a sealed container holds them under these conditions while in space. When the seam broke, so did the electronics.

The spacecraft should have passed by Venus on 19 July but by that point it had stopped communicating and Jodrell Bank could detect no sign of the spacecraft.

Never one to let a trio of catastrophic failures get them down, the Soviet space programme pressed on. Four missions were planned for the next launch window in autumn 1965. Two of these would fail; two would not.

The second mission to earn itself a Venera designation, Venera 2, was a flyby mission. When it left Earth on 12 November 1965, its exit from parking orbit was so accurate that even before its mid-course correction it would fly past Venus at a distance of 23,950km, 16,000km closer than expected.

It was all going well – too well.

As the spacecraft neared the planet, it reoriented itself to give the planet its sole focus. To do that, it had to stop communicating with Earth. Instead, it would record all the data from the mission, transmitting it after the flyby. On 27 February 1966, it cut communications to prepare to pass the planet, never to be heard from again.

Perhaps Venera 3 would have more success. It had left Earth just four days after Venera 2, and this time the plan wasn't just to pass Venus by, but to hit it. The initial plan was for a lander to do this, but it too lost contact with Earth. Although it wasn't able to communicate, the spacecraft was already on a collision course with Venus. It seemed that *all* of Venera 3 would end up being the lander.

The probe crashed on 1 March. There was a good chance the descent probe had detached and landed automatically, but even if it had still been attached to the mothercraft when it crashed the result was the same – Soviet hardware had made it onto Venus.

Looking at the last transmissions from both spacecraft, it seemed the probe's radiator was the cause of their dual failure. The team redesigned the large radio dish from the high-gain antenna – a directional radio dish that is the main communications array on a spacecraft – to redistribute the heat and keep things cool.

Finding out about these kinds of flaws mid-flight isn't the most efficient way to design a spacecraft. By January 1967, a facility was built at Lavochkin where spaceflight engineers could test their spacecraft before they flew. It quickly proved its worth. When testing the landers (which had yet to have the chance to actually land), the Venera team found that they stood no hope of surviving the extreme temperatures and G-forces they'd experience descending through the atmosphere. They were redesigned to endure temperatures of 11,000°C and up to 300G. Now they just had to get that far.

Another three Venus missions were attempted in the June 1967 window. One of these was Mariner 5, a flyby mission from NASA that flew past on 19 October that year. The other two were Soviet endeavours. The second of these failed, but the first successfully launched on 12 June, becoming Venera 4.

Everything went well during its five-month journey. Things were looking good for a Venus landing on 18 October. As the morning arrived, Jodrell Bank listened in. The radio played the steady signal of the mothercraft until it suddenly fell silent, only for a new broadcast to sing out from the planet. Had the Soviet's finally landed on another world?

Not exactly. Venera 4 had released its lander 44,800km above the surface. The probe slammed into the upper atmosphere, subjecting the craft to 450G as it rapidly slowed down. When the probe determined it had slowed enough, it released its parachute.

The further it fell towards the alien planet the worse conditions grew. At first, the temperature was a balmy 31°C, but quickly grew hotter. And hotter. And hotter. As the temperature rose, so did the pressure as the air grew increasingly thick. The probe held on far as long as it could, but after 93 minutes the conditions became too much. Venera 4's last measurements showed an air pressure twenty-two times that of Earth's and temperature of 271°C.

At first, the Venera team thought the lander had stopped working when it crashed into the surface, but the last altimeter reading indicated it was over 20km above the ground. If these were the conditions that high up, what was it like on the surface? Combined with Mariner 5 readings, it was becoming apparent that temperatures and pressures on Venus were higher than anticipated – much, much higher.

The hopes the Soviets harboured of sending a human to the surface of Venus were now firmly quashed. Any spacecraft they could conceivably build would eventually succumb to the climate, killing any humans on board. In fact, they'd be lucky to survive more than a few hours. Fortunately, no one cares if you send robots to their certain death, and so the Venera programme continued.

While the spacecraft had been built to withstand 11,000°C, the cables that held the parachute in place were designed to withstand only 450°C. Extrapolating the temperatures further down the atmosphere, it's almost certain the cables melted through long before reaching the surface, dropping the lander from several kilometres up. That was, if the main cabin hadn't cracked under the pressure. Given the unexpectedly extreme conditions, even getting that far was an impressive achievement.

Venera 4 hadn't lasted long, but it had hinted at exactly what sort of planet Venus was. The probe found the atmosphere was predominantly carbon dioxide, with just under 10 per cent nitrogen; experts had expected the ratio to be the other way around. The probe had found a halo of hydrogen around the planet, suggesting there was water in the atmosphere being broken apart by the Sun's radiation even though it was far too hot for any water to exist on the surface.

This time the failure of the mission came not from a faulty rocket or probe, but from the Venera team underestimating just how extreme the planet was. The entire Venera landing capsule had to be redesigned, from the cabin itself to the parachute that carried it down. To survive the descent, the probes needed to get to the ground as fast as possible before the capsule was crushed by pressure or the parachute melted. That meant a smaller parachute. Meanwhile, the lander cabin was strengthened and fitted with an improved altimeter, allowing them to measure their altitude much more precisely if they once again failed to reach the surface.

In January 1969, both Venus probes got under way successfully. Even better, both Venera 5 and 6 survived the journey. At least the Soviets had ironed out the issues of the transit stage of the spacecraft. Now it was up to the capsules to survive the descent.

Venera 5 spent 53 minutes descending through the atmosphere before succumbing to the heat after 37km of falling, while Venera 6 travelled 34km in 50 minutes before expiring. Although they didn't quite stick the landing, it's hard to call Venera 5 and 6 failures given what they had to endure. Their findings backed Venera 4's and if the trend of pressure and temperature rise continued, then surface pressure could be over 100 times that of Earth with temperatures over 400°C.

The Soviets had spent decades dreaming of travelling to other worlds, visiting Venus and finding life beneath its clouds. Although they had managed the journey, it was hard to believe that anything, even bacteria, could survive in such a harsh landscape. People had dreamed of finding a garden world beneath Venus's clouds. Instead they found a hellscape.

8

ON THE SURFACE OF VENUS

The discovery of Venus's true nature didn't sate the Soviets' appetite for exploration. They had managed to descend through the planet's skies; now they sought to land on its surface. Getting there in one (uncrushed) piece would require a very specialised lander. The Venera team called upon the experts in surviving under pressure – submarine engineers.

Their suggestions were invaluable but created new challenges. The cabins needed to be crafted from a single structure, a perfect sphere with no welds or holes that would give way under pressure, a manufacturing nightmare. To keep the weight down, the entire structure would have to be made from lightweight titanium, a metal that had only recently become usable, meaning the machinists had no experience working with it. The end result was a bulky probe that had little room for scientific instrumentation. But the cabin could now withstand up to 180 atmospheres of pressure and temperatures of 540°C. Even the worst estimates of the surface conditions weren't that bad.

The aim was to get at least 90 minutes out of the probe. That would be enough time to descend through Venus's atmosphere, reach the surface and send back confirmation. To keep the descent time down, the probe used the heat to its advantage. In an ingenious piece of design, Venera's parachutes now had cables that stopped them opening fully at first. This kept the probe from slowing down too much, so it quickly descended through the atmosphere. As the temperatures rose over 200°C these cables would melt through, allowing the parachute to fully open and slow the lander to a survivable speed.

By August 1970 launch window, the new breed of landers was ready. Again, a pair was sent, only one of which made it far enough to become Venera 7.

It reached Venus on 15 December 1970 and began to make its descent. As it sank deeper into the atmosphere, the temperature and pressure began to climb. Just as planned, the cables melted through. The parachute bellowed out in the Venusian air and then … it tore! The parachute swung wildly and within a few minutes the fabric collapsed. The lander plunged to the surface, striking at a speed of 17m/s (about 60km/h).

Venera 7 survived its ignoble landing. The stocky probe had truly been built to withstand anything. However, the lander was on its side, meaning its main transmitter was pointing in the wrong direction as it called out for home. Only a tiny fraction of the transmission made it to Earth, but radar technician Oleg Rzhiga managed to strain his ears enough to hear the faint signal coming in.

Venera 7 held out for 23 minutes, revealing that the surface pressure was 92 atmospheres, about the same as you'd encounter 920m underwater. Meanwhile, the temperature was 474°C, hot enough to melt lead.

It hadn't been an elegant landing. But a landing it undoubtedly was. Venera 7 had proven it was possible to reach the surface of Venus in one piece. The Soviets just needed to fix the parachute.

Knowing the probe only had to withstand 92 atmospheres rather than 180, the lander could lose some of its anti-crush hardware, allowing more room for scientific instruments. When Venera 8 began its way to Venus on 27 March 1972, it was loaded up with experiments to measure the light levels on the surface, wind-speed measurement devices, temperature and pressure sensors, gas analysers and even a gamma ray spectrometer to test the composition of the soil.

Venera 8 headed off on its way, and arrived at Venus on 22 July, aiming for the daylight side of the planet. It hit the atmosphere with a speed of 11km/s. By the time it reached an altitude of 67km, it was going just 0.25km/s.

At 60km up, the parachute opened.

At 50km, the instruments turned on. They began to measure the atmosphere as the probe travelled down towards the surface.

At 45km up the spacecraft began to take radar readings, creating a map of the ground that lay beneath.

By 30km, the first parachute wires melted through, allowing the parachute to open fully.

After just over 50 minutes travelling through the atmosphere, Venera 8 touched down; softly this time. On Earth, the signal came in loud and clear, its antenna pointing straight and true. The spacecraft transmitted for 63 minutes.

The sensors showed that, surprisingly, there was very little difference in temperature between the day side it had landed on and the night side Venera 7 had measured. The photometers found that the light levels gradually dropped between 50–35km, indicating clouds, but from 32km down the air seemed to clear. On the surface, the light levels were akin to Earth around sunrise.

During the descent, the probe's chemical analysers found the clouds were made of sulphuric acid. As well as being another deadly attribute to add to the killer planet's make-up, it answered the question of how you form 'dry' clouds on a planet with no water.

Planetary scientists back on Earth now had a good idea of what your typical morning on Venus was like. It was permanently overcast. The pressure was enough to crush you, the heat could boil you alive and the atmosphere might make you dissolve. But at least there was a nice, gentle breeze …

Venera 8 was a momentous achievement for the 3MV line, but it marked the limit of what the decade-old series could do. With a fuller understanding of what Venus would throw at them, the Soviets went back to the drawing board to design a new, better machine specifically designed for exploring Venus – the 4V-1.

One of the biggest innovations was a major change in how the entire mission was conducted. From now on, rather than flying off into deep space, the main spacecraft would stay in orbit around the planet. As well as being able to study Venus in detail from above, the orbiting spacecraft could now act as a communications relay for the lander.

Previously, the Venera landers had communicated directly with Earth. However, the lander could only carry a small antenna and battery, meaning it had to reduce the amount of data it was sending back to ensure the signal was clearly readable once it reached Earth.

For the 4V-1 missions, the orbiting spacecraft would act as a relay station, with the lander transmitting to the orbiter, which then passed the signal on to Earth. The orbiter could have a much larger radio dish, a beefier battery and be free to take its time transmitting the data home. Combined with a decade of technological advances, the amount of data the spacecraft could transmit tripled.

The lander was redesigned to approach at a shallower angle, so it would only have to withstand around 170G, rather than the 400G the 3MV had experienced. After the problems with the 3MV's parachute, the 4V-1 line used a new approach during descent. The parachute would cut away at 50km and a novel ring-braking system would use the thick atmosphere to slow down to a survivable 7m/s. This would reduce the descent to 20 minutes, giving more time on the surface. To maximise this even further, the instruments would be pre-cooled while attached to the orbiter, meaning they could last longer before they overheated.

The science instruments themselves were far more ambitious. As well as updated versions of the previous experiments, there was equipment to measure radiation, the density of the surface, and a mass spectrometer to give a much boarder elemental analysis of the atmosphere.

One of the most important additions was a pair of cameras looking out through special windows cut into opposite sides of the lander. Together, they could conduct a 360° survey. It was a slow process, taking 30 minutes to create one panorama and transmit it back to the orbiter but the new lander should easily survive that long.

It took several years to design and build the new spacecraft, but when the window in 1975 rolled around, two 4V-1 spacecraft were ready to go. They both launched successfully in June 1975, beginning what would go on to become a golden age of Venus exploration.

Venera 9 arrived at the planet on 20 October, dropping off its lander as the main spacecraft passed into orbit. Two days later, the probe hit the atmosphere and deployed its parachute. This cut away at 50km, just as intended, at which point the ring brake kicked in, using the atmosphere to slow down. It worked perfectly. All the while it transmitted data back to the orbiter, ready for transmission back to Earth later.

The lander came down 2.5km above 'sea level' in a region known as Beta Regio on the slope of either a hill or a crater – it was hard to tell. Knowing that time on the surface was limited, the team ordered the spacecraft to begin its surface operations as soon as it touched down. After just 2 minutes, the spacecraft readied its most anticipated experiment – the cameras.

The lens caps popped off, but unfortunately the air pressure stopped one from ejecting, while the other came away cleanly. A light gauge measured

ВЕНЕРА-9 22.10.1975   ОБРАБОТКА ИППИ АН СССР 28.2.1976

Venera 9's view of the Venusian surface. The camera makes the image appear curved,
but the horizon is just visible in the corner of the image. (NASA/ NSSDC/ Russian
Space Agency: nssdc.gsfc.nasa.gov/imgcat/html/object_page/v09_lander.html)

how bright the conditions were – although Venera 9 had landed on the
day side, the thick atmosphere meant it was so gloomy the lander had to
use its on-board floodlights.

After half an hour, Venera 9 started to transmit its first panorama. Back
on Earth, people weren't very hopeful of getting a good image – the
planet is dusty, dark and cloudy, none of which are helpful when you are
trying to take a decent photograph. However, the image they received
was remarkably clear. The view was clear to the horizon 300m away, and
the air seemed free of dust. The surrounding was littered with rocks, their
edges sharp and angular with no signs of wind erosion.

The next day, sister probe Venera 10 bumped down in Beta Regio. Once
again, one of the cameras failed, preventing a full panorama, but the images
that were sent back revealed a vista with far more weathering than Venera
9 had seen, suggesting the surface was much older.

With both missions a success on the first try, things were looking up for
the Soviet planetary exploration programme. The Venera team decided to
forgo the 1977 launch window to upgrade the next round of landers to
perform a more detailed chemical analysis. They had a drill that would be
able to transfer material to an on-board laboratory. Here, a spectrometer
was able to measure what it was made of. The transmitters had also been
upgraded, increasing their bandwidth twelve times.

The spacecraft launched late in the 1978 window, so they had to travel
slightly further, meaning they didn't have as much fuel when they reached
Venus. The orbiters couldn't slow down enough to enter orbit – they were
destined to be flyby missions, after all. Fortunately, the orbiters stayed in

sight of the landers long enough to radio home all their data. In fact, they remained in contact longer than if they had entered orbit.

Due to a quirk in the orbital mechanics, Venera 12 arrived first on 14 December, despite launching second, entering the atmosphere seven days later. The lander managed to tough it out for 110 minutes on the surface, the longest of any Venera mission. It only dropped out of touch when the orbiter moved out of range.

Venera 11 arrived on Christmas Day* and lasted 90 minutes, again only stopping when the spacecraft passed out of range. Both had come down to the east of a region called Navak Planitia.

With the landers down, the world waited for the images. The signals had been recorded in the West, but the Soviet transmissions had been encrypted since the Luna 9 affair. The Soviets announced that they had panoramas from the landers, showing a flat, rocky landscape, but didn't release the pictures. Then the story changed, stating that the probes had come down at night and they had no way to take images, another lie as they had floodlights.

The truth was, the Soviets had taken the ambitious step of including colour cameras. The camera would take four images through different filters (red, green, blue and clear), then stitch these together, using a calibration chart to make sure the colour balance was accurate. However, this increased each image size fourfold, hence the increase in the radio's bandwidth.

The controllers waited for the colour images to come in, but when the long-awaited transmission came back it carried no data. This time, the pressure had prevented all four lens caps ejecting, completely blinding them. To make matters worse, a new system to drill and analyse soil samples on board failed, as did the instrument meant to measure the density of the soil. The two landers might have spent the longest time on the surface of any Venera mission, but it was time spent with their arms tied behind their back.

Despite the failures, the mission returned a huge amount of data, although mostly from the descent stage. A new gas analyser sniffed the air, finding traces of argon, which is usually created by the radioactive decay of potassium and signifies past volcanism in the area. In the high atmosphere,

---

*   Religion was widely suppressed in the Soviet Union in the 1970s, seeing it as the 'opiate of the masses'. Those who did maintain a religion were largely Russian Orthodox Christians, who celebrate Christmas in January. Either way, no one missed their *pirozhki*.

the probes detected sulphur and chlorine, emphasising what a thoroughly unpleasant place Venus was to visit.

Despite the lack of cameras, sensors still managed to detect how much light was getting through the clouds – only around 5 per cent. What did make it through was dispersed by clouds so thick you wouldn't even be able to tell which direction the Sun was in the Venusian sky.

However, perhaps the most intriguing result came not from an instrument trying to touch Venus, sniff it or look at it, but one listening to it. The Groza experiment was a microphone listening for the rumble of distant thunderstorms.

Venus, it turned out, is indeed a dark and stormy planet. Venera 11 measured as many as thirty lightning strikes in a single second, while Venera 12 heard over 1,200 during its journey to the surface. As Venera 12 set down, it was heralded in by a thunderclap so enormous it rang across the planet for 15 minutes. Fortunately for the probes, these strikes stayed within the clouds and never seemed to hit the ground.

The Soviets had successfully reached the surface of Venus with Venera 7 and taken a look around with Venera 9, but the extensive problems suffered by Venera 11 and 12 required a re-examination of their probes. For the next few years, the Soviets overhauled the 4V-1 design to ensure the next batch of probes wouldn't suffer the same humiliating failures. They would miss the next launch window, but it was better to have a late, successful mission than an on time, lame one.

9

THE UNITED STATES AT VENUS

Venera 11 and 12 were not the only probes to grace the Venusian sky in 1978. The United States had not forgotten about the planet. In the years when Apollo had been getting all the attention, the Pioneer programme had been exploring the deep solar system. Pioneers 10 and 11 were at that moment streaking outwards, having flown past Jupiter in 1973 and 1974 respectively. With the end of the Apollo programme in 1972, the agency began to look beyond the Moon and prepared to send out its first probes to our neighbour planets.

For Venus, NASA designed a pair of spacecraft, the Pioneer Venus Orbiter and the Pioneer Venus Multiprobe,* using the results published by the Soviets. The orbiter would map out the planet from above.

The Multiprobe was in fact five different spacecraft in one, all of which were destined for a date with Venus's atmosphere. The biggest component of the Multiprobe was the 315kg Large Probe. Like the Venera landers, this was equipped with a heat shield and a parachute that would help guide the probe down through the atmosphere. As well as instruments to measure the atmosphere's composition, it carried light sensors that looked out at the planet through windows made of diamond – the only material tough enough to withstand the pressure.

---

* As things can never be named straightforwardly: The Pioneer Venus Orbiter was also known as Pioneer Venus 1 or Pioneer 12. Meanwhile, the Pioneer Venus Multiprobe was known by Pioneer Venus 2 and Pioneer 13. I've chosen to call them Orbiter and Multiprobe because in this context it makes keeping track of them much easier.

Alongside this were three spheres the size of a beach ball, each weighing 90kg. These had no parachutes, although they did have protective aeroshells. They carried a much simpler set of instruments to measure the temperature and pressure of the atmosphere, but also a radio to track their motion as they were buffeted about by the wind.

Finally, there was the bus – the carthorse of the Multiprobe, whose main purpose was to carry the other four probes, although it did contain a few instruments to measure the composition of the atmosphere when it too hit Venus.

None of the probes were ever designed to land on the surface. In the years following Apollo, NASA found their financial belts significantly tightened. Leading up to the Pioneer Venus missions, NASA were prioritising the Voyager probes, which would take advantage of a once in a lifetime alignment of the outer planets. The Soviets had shown landing on Venus was hard, which made it expensive, so the requirement was scrubbed from Pioneer Venus's plans. Instead, the aim was to investigate the atmosphere. If a probe happened to survive for a while on the surface, so much the better.

The quintet launched on 8 August 1978, arriving at Venus 123 days later. The Large Probe was the first to splinter off, on 16 November. Four days later, the Small Probes peeled off one by one. As they flew through the void, their staggered detachment caused them to pull apart.

All the probes hit the atmosphere on 9 December. The first to arrive was the Large Probe. The Small Probes soon followed, thudding into different locations on the planet, earning themselves the new names North, Day and Night, depending on where they struck. The Bus was the last to strike. As it wasn't designed to survive the descent, it broke up 110km above the surface. The Large Probe transmitted throughout its descent, but the parachute cut away 47km above. With no other method of slowing down, its transmission abruptly stopped when it smashed into the ground. Of the small probes, only Day survived beyond its abrupt meeting with the surface, lasting a further 67 minutes.

By and large, the results backed up the work conducted by the Soviets. However, the probes did find that the signature of argon in the atmosphere was different to that of Earth's. This factor suggested the two planets would be fundamentally different even if Venus had not undergone the runaway greenhouse effect that had resulted in the planet's hellish atmosphere.

Around an altitude of 12.5km, all the spacecraft experienced strange electrical discharges that interfered with their communications and broke several sensors. To this day, no one is sure exactly what caused what has become known as the '12.5km anomaly', although it does seem certain that some quirk of Venus's atmosphere is to blame. Because of the Soviet policy of painting everything to be sunshine and roses, the United States weren't sure if the fault lay in their design or if the Soviets had suffered a similar problem.

Despite these issues, the mission was largely a success. However, they were also the last NASA would send to Venus's surface. The planet was a difficult one to explore. Not only did the crushing atmosphere make exploring it from below extremely challenging, the thick clouds made looking at it from above difficult too. There was only so much money, and there were other planets offering far more attractive, not to mention easier, places for the agency to explore. Perhaps it would be best to leave Venus for the Soviets.

10

TOGETHER AT VENUS

By 1981, the Cold War tensions that governed so much of the early Space Age were beginning to ease, allowing a far greater degree of international collaboration. By working together, rather than at odds, the Soviets hoped to push their missions further than they would have been able to alone.

Just as the Pioneer Venus probes used Soviet data, the Venera teams worked with US geologists and the Pioneer Venus Orbiter maps to pick out the best places for their next set of probes. The lens cap issues of the 4V-1 line had now been ironed out, and on 30 October, Venera 13 began its journey towards an ancient-looking terrain. A few days later, on 4 November Venera 14 followed, heading for what seemed a younger surface, recently covered over by volcanic activity.

Venera 13 landed on 1 March 1982. The landing was harder than intended, but vicious metal teeth on the probe's underside bit into the rock, holding it secure. There was no tense wait to see if the lens cap would stick – it popped off shortly after landing. With limited time before the probe succumbed to the pressure, and each colour image requiring four separate exposures, the camera got straight to work. Within minutes it took a full panorama with all four filters and sent them back, ready to create the first colour image from another world.

The first, hasty picture revealed a truly alien-looking world, where yellow-green rocks lay under an orange-tinted sky. The landscape was stony, covered with great cracks half-filled with dust, while the larger rocks looked like they had been worn away by the slow beating of the wind – this did indeed look like an ancient part of Venus. With the first 'insurance' image taken, Venera 13 took its leisure with the next to gain more detail.

During this photo shoot, the rest of the probe had not been idle. The drill began work shortly after landing, and within 4 minutes made its first analysis. Over the next 32 minutes, the spacecraft methodically took its samples as the heat and pressure bore down, revealing the rocks to be volcanic, as expected. Yet it could not hold out forever; 127 minutes after landing, Venera 13 abruptly stopped.

Venera 14 landed four days later and 1,000km away. Here the landscape was smooth and flat to the horizon, and relatively free of rubble – all hallmarks of a young surface, a hypothesis that was backed up by the chemical analysis.

One experiment that didn't go quite to plan was the spacecraft's penetrometer, which was to measure the density of the topsoil. It seemed the lens cap was once again at fault. Although it came off just as it was meant to, it fell exactly where Venera 14's probing arm meant to deploy the penetrometer. Fortunately, Venera 13's worked perfectly – another demonstration of the wisdom of doubling up space missions. Venera 14 could not match its sister's longevity, though, and failed after just 57 minutes.

The next two spacecraft, Venera 15 and 16, were orbiters, replacing the lander with a radar system that could pierce the clouds to map the landscape below. They would also be the last. For twenty years the Venera programme had been an enormous success, revealing the planet to be an incredible, if horrific place. But the programme had run its course. It was time to take everything they'd taught us about the planet and come up with a new way to explore Venus.

While the orbiters could provide a wide-scale view of the planet, that view lacked detail. The landers, meanwhile, gave an immense amount of detail, but only of the narrow range of spots where they landed. What was needed now was something in between the two.

The caustic environment meant human exploration was right out and a rover wasn't any more feasible. Instead, would-be Venus explorers turned their eyes to the sky. The thick atmosphere that had been the source of countless problems could be an asset. Rather than go to the surface, why not float across the air? The Soviets had been contemplating such an endeavour alongside a team of French scientists; a collaboration that began in the most unlikely of places – a cocktail party.

While at a soiree in the late 1970s, Jacques Blamont, a space pioneer who helped set up the French space agency, chatted with Soviet scientists about an idea he'd had to explore Venus's atmosphere by suspending a probe on the bottom of an air balloon. If they could get the mission ready in time, there was even the chance to take a small detour to meet a rare celestial visitor – Halley's Comet. One of the solar system's most famous comets, Halley was due to pass by Venus in 1985. If the mothership flew past Venus rather than staying in orbit, it could easily be sent towards the comet, offering two space missions for the price of one.

It was too good an opportunity to pass up. A plan discussed over drinks became a mission called Vega, an amalgamation of the Russian for Venus and Halley. The Soviets sent out an international call for other nations wanting to fly their instruments on their spacecraft. Thirteen nations responded.

Rather than simply building elements and sending them on to the Soviets, these foreign scientists and engineers were invited to Moscow to work alongside the mission planners during all stages of construction. The Czechs would build the telescope to follow Halley's Comet, the navigation would be built by Hungary and the French would help support with communications. Even the United States took part in the project, albeit via the French because Cold War tensions were still too hot to work together directly.

Two Vega spacecraft launched on 15 and 21 December 1984, arriving at Venus in June 1985. They carried with them a traditional Venera-style lander. Vega 1 suffered from the same 12.5km anomaly as the Pioneer Venus, causing it to start its landing sequence while still in the air, and the mission was lost.

Luckily, Vega 2 landed as planned. It had dispensed with cameras in favour of experiments to measure the atmospheric and surface composition.* It discovered that the rocks it had landed on appeared to have formed in a wet climate, hinting that Venus may not always have been as dry as it is now.

---

*   In a bizarre turn of events, the Soviet media at first claimed that they couldn't take images because it was night, rather than the perfectly reasonable truth that they hadn't thought cameras to be scientifically necessary. It was an odd occurrence for a propaganda machine that usually tried to cover over failures, rather than create them.

The jewels of the Vega missions were the balloons. Dropped into the atmosphere on the planet's night side, the balloons inflated to their 3.5m diameter. But, because nothing on Venus can be pleasant, instead of bobbing along calmly, the balloons were immediately thrown into hurricane-like vortices. They were sitting around 54km above the surface in one of the planet's most extreme layers. Winds reaching 240km/h buffeted them from all sides. Air pockets suddenly opened up beneath them, making the balloons plummet 2km in a few seconds, then moments later, a gust of wind would throw them back up in the air again.

Despite this rattling ride, both probes lasted almost two days. Vega 1 was buffeted around 9,000km through the Venusian atmosphere before its battery died after 46.5 hours. Vega 2 suffered a similar fate. In time, the balloons would have floated around to the day side, where the Sun's heat would cause the balloons to expand and burst, sending the instruments suspended below on a terminal journey to the surface of Venus.

The Vega missions were a success not just for the Soviets, but for Austria, Bulgaria, Czechoslovakia, France, the Federal Republic of Germany, the German Democratic Republic, Hungary and Poland, all of whom helped take part in the mission, highlighting what could be achieved if nations worked together.

These achievements would no longer be at Venus, however. To date, neither the Soviets nor any other nation has attempted to return to Venus's surface. When they began to explore Venus, the Soviet Union hoped to find a second Eden. Instead, they found hell. A thick blanket of carbon dioxide boiled the planet alive, while clouds of sulphuric acid floated through the skies. Lightning strikes tore through the upper atmosphere, their thunderclaps loud enough to be heard around the planet, while storms more violent than Earth's worst hurricanes blew through the upper atmosphere. At the surface, the pressure was a crushing 90 atmospheres and temperatures averaged 464°C. Even the soil contained high levels of radioactive uranium.

While there were still many questions left about the planet (such as, what was it about Venus's past that had caused it to veer off to create such a violent world compared to our own?) the Venera and Vega missions had given planetary scientists a rich enough picture of the world to begin the search for answers. But they had been pushed as far as they could go. In 1985, the Soviets decided to refocus their efforts elsewhere.

This was a disastrous decision, at least for the legacy of Soviet planetary exploration. Because of the Soviet propaganda machine's love of secrecy, many in the West forgot about the hugely successful Venus missions. Instead, the lasting memory of Soviet planetary exploration (at least in the Western mind) would be the string of failures at Mars.

PART 3
MARS

11

THE CURSE OF MARS

Although the red pin prick of light known as Mars has been seen wandering around the sky since ancient times, humanity's love affair with the planet truly began in 1877, when Italian astronomer Giovanni Schiaparelli claimed he'd seen '*canali*' on the surface. A few decades later, US astronomer Percival Lowell not only claimed to have seen them, but he also made extensive maps of their structure. The idea quickly gained popularity with the general public, despite the fact that no other astronomer could find any trace of these 'canals'.*

Thinking Mars might be home to some highly intelligent alien civilisation, the planet became the setting for several enormously popular works of science fiction, such as the novel *Aelita* by Aleksey Tolstoy** and the *John Carter* novels by Edgar Rice Burroughs, both of which involved Earth astronauts seducing beautiful Martian princesses. H.G. Wells then brought the Martians to Earth in his *War of the Worlds*.

In more realistic terms, Wernher von Braun wrote *Das Marsprojekt*, a book about how humanity might travel to Mars – an epic mission involving multiple ships each staffed by up to seventy astronauts.*** It was first published in English in 1953, before being carried to a wider audience in

---

*   '*Canali*' is actually Italian for 'channels', a rather more mundane term than the intelligently crafted 'canal' translation. Schiaparelli himself did little to correct the mistranslation.

**   No, not that Tolstoy.

***   Von Braun's initial plan consisted of sending multiple ships to Mars, based on the premise that not all of them would make it. While pragmatic, this wasn't a very American way of doing things. Begrudgingly, he changed his ideas.

1957 when von Braun discussed his ideas on the Disneyland episode, *Mars and Beyond*.

In the early 1960s, it quickly became apparent that Venus wasn't hiding a lush rainforest beneath its clouds, meaning that Mars now shouldered the hopes of those dreaming of finding life beyond Earth. But first, they'd have to get there, and that would prove no easy feat.

Mars has a reputation for being a cursed planet – a reputation that is well earned. Of all the missions to Mars, half have failed to reach the planet. While such a high failure rate was not out of place during the early days of the Space Age, Mars's fatality count has failed to fall and has claimed the lives of robotic explorers throughout the decades.

In many respects, the planet is no harder to reach than Venus, as it's about the same distance away. Those missions that failed in transit were largely down to bad luck. The trickiest part comes when you try to land on Mars.

On the Moon, the gravity is low enough that landers can slowly fall towards the surface, and retrorockets can easily counteract the pull of its weak gravity. But Mars is much larger, and its gravitational pull accelerates spacecraft towards it far too fast for retrorockets alone to counter.

Meanwhile, Venus has a thick atmosphere that spacecraft can harness to slow down, either with heat shields or parachutes. Mars's atmosphere, meanwhile, is only 1 per cent of Earth's surface pressure – too thin for all but the largest parachutes, but too thick to ignore. The Red Planet has the worst of both worlds – a reverse Goldilocks planet, where everything is just wrong.

Of the many failures to reach Mars, a huge number belonged to the Soviets. The twenty missions that the Soviet Union tried to send to Mars almost all ended in failure. Not one fully completed its mission.

The early missions were based on the same 1MV spacecraft that attempted to go to Venus. No one expected these to reach Mars, but they could investigate what the radiation and magnetic environment was like in interplanetary space, as well as testing out the technology, meaning the next run of probes would be more successful. None of them made it off the launch pad.

Launch windows to Mars come around once every twenty-six months, so the Soviets had to wait until 1962 for the next chance. The agency took the opportunity to launch three of the new 2MV probes towards the planet, the last of which carried a lander to look at the life they hoped was thriving on the Red Planet.

The first failed to even leave Earth, exploding during launch – an event that occurred during the Cuban Missile crisis. The explosion triggered US alarms, but a quick analysis of the radar showed the true cause. Thankfully.

The probe carrying the lander made it to orbit, but no further. It ended up crashing into the Kazakhstan desert. When Korolev sent a team to recover the probe, he ordered them to deploy the experiment that had been meant to detect life-signs on Mars. Despite there being abundant life at the crash site, the instrument found none, so it was removed from future probes.

One of the three did make it under way and became Mars 1. Jodrell Bank followed the spacecraft as it made its way across the heavens towards Mars. At first the probe seemed healthy, but it soon developed a leak in one of the tanks containing the propellant gas it used to adjust its course. The leak sent the spacecraft spinning and with no fuel to compensate, Mars 1 slowly drifted further out into space.

The probe finally fell silent when it was 106 million km from Earth. It would have passed by Mars on 19 June 1963, albeit at a distance of 193,000km rather than the 1,000km it was meant to pass at.

In 1964, the Soviets managed to get one Mars probe into space, only for the solar panels not to deploy owing to a design fault. With no time to rectify the issue, the following missions were cancelled. The United States flew past the planet with Mariner 4 on 14–15 July 1965, a rare victory for the United States during the early phase of the Space Race.

During the next flight window in 1969, while the United States were preoccupied with landing Armstrong on the Moon, the Soviets planned the Mars 69 mission. By now, the engineers knew enough about the Red Planet to develop a specialist probe, rather than relying on the generic 3MV being used at Venus. Unfortunately, they never got to test their new design, as the mission fell victim to the persistently problematic Proton rocket that had claimed so many of their would-be lunar missions.

In 1971, the probes went through yet another redesign. The plan was now to send an orbiter first to scout the way, followed by two soft landers. The surface probes would be equipped with cameras, weather stations and soil examiners and carry a small rover known as PrOP-M, joined to the lander by a 15m cable.

The United States also planned to send two orbiters, Mariner 8 and 9, so the race was on to see who would achieve orbit first. Initially, things looked up for the Soviets when Mariner 8 crashed into the Atlantic Ocean. Unfortunately, their own orbiter followed suit two days later.

The two Soviet spacecraft bearing landers made it under way in May 1971, now called Mars 2 and 3. However, the transmitters on both were having problems. To make matters worse, NASA's Mariner 9 launched successfully, with a trajectory that meant it would beat the Soviets to orbit. Then, Chief Designer Georgi Babakin, the man who had succeeded Korolev, died on 3 August. The Soviets could only hope their streak of bad luck was broken by the time the landings came around.

As the spacecraft approached Mars, the Red Planet's weather forecast came in – a storm of red dust was encircling the planet. These maelstroms occur on Mars once every two to three Martian years (a Martian year is a little less than two Earth years). While the orbiters would remain above the storm, the landers didn't have such luxury. They would have to come down in the middle of it.

Mars 2 arrived at the planet on 27 November. An electrical fault released the lander along the wrong path, and it hit Mars's atmosphere at too steep an angle. It crashed to the surface without time for its parachutes to open. Mars 2 did become the first human-created object to land on Mars but not in the way the Soviets had wanted. They could only hope that Mars 3, following on 1 million km behind, would fare better.

On 2 December 1971, Mars 3 detached from the orbiter and began its descent. This time the probe took the correct trajectory, and the spacecraft successfully touched down on Mars. The Soviets had done it. The first soft landing on the Red Planet had been completed by a 'Red' spacecraft.

Exactly how soft the landing was is debatable. According to Americans spying on the landing, the probe came down at a rather speedy 20.7m/s. But, crash landing or not, the probe was transmitting from the surface. Back in Russia, the Mars 3 team – now led by Sergei Kryukov, a quiet man who was still getting used to his sudden promotion after Babakin's death – were patiently waiting for the lander to unfold and begin imaging the surrounding terrain.

After 2 minutes on the surface, Mars 3 began to send back the first ever images from Mars. And it transmitted them for a whole 14.5 seconds before the signal disappeared. Through the static and noise, it was just possible to make out the line between the bright sky and the dark horizon in the partial picture that managed to get through. But the low quality of the image meant it was soon resigned to the archives of the Vernadsky Institute.

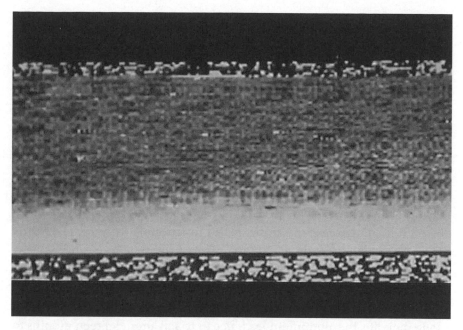

The first 'image' from Mars's surface. The transmission cut out after just a few seconds, leaving little more than noise. (Russian Space Agency)

Several years later, researchers unearthed the image and printed it out at a higher resolution and found the shapes of boulders littered across the few inches of surface that Mars 3 had managed to transmit. It was a tantalising hint of what the mission had almost achieved, only to fail at the last minute. The exact cause of the signal loss was never discovered but the blame is usually laid on the dust storm raging at the time.

Despite the almost success, the Soviet plans to invade Mars proceeded unabated. Four Mars missions were planned for the 1973 window based on an updated design of the Mars 2 and 3 probes – a pair of orbiters and a pair of landers.

The class of 1973 was probably the Soviet's most successful set of Martian missions. They all made it out of orbit and received the Mars moniker, with the two landers numbered 6 and 7. Unfortunately, that is pretty much the only way the spacecraft were a success.

A few months before launch, a check on the lander picked up a fault. The technicians gathered around to diagnose the problem – a 2T-312 transistor. These are a key part of any electronic circuit. The devices act as switches and amplifiers, controlling the flow of electricity. If they break down, then your electronics stop working.

Transistors are often replaced between successive iterations of not just spacecraft, but all robotics and electronics, due to something called Moore's Law. In 1965, CEO of Intel Gordon Moore noted that developments in transistor technology meant that every two years, the number of transistors you could fit in a given space doubled, while the cost halved. In Earthbound computers smaller transistors translated to more powerful processors capable of performing an ever-increasing number of calculations. In spacecraft, upgrading transistors translated to reducing the weight and size of electronic hardware.

Changing components on a working spacecraft is always a risky endeavour, especially when you have to buy those components from someone who doesn't fully understand the challenges of functioning in space. The Soviet Mars team procured their 2T-312 transistors from the Voronezhskiy plant. Previous transistors were manufactured with gold leads but some enterprising soul trying to save a few rubbles had replaced them with aluminium.

While gold is largely unreactive, giving electronics manufactured with it a long shelf life, the aluminium leads began to fail after eighteen months to two years – about the time it takes to build, launch and get a spacecraft to Mars. While the manufacturers at Voronezhskiy had ensured the aluminium transistors would work as well as the gold ones, no one had checked to see if they would work for as long.

Each spacecraft was filled with hundreds of transistors that were now beginning to fail. Flying with them in place would have a 50 per cent chance of failure, but replacing them would take six months. They'd miss the launch window. To make matters worse, many of the transistors had been installed backwards on Mars 6, causing the probe to be seriously damaged during testing. The technicians reinstalled the transistors, but there was no time to fix any additional damage, leaving them with a very sick robot.

While under other circumstances, the Soviets might have decided to take the time, the United States were also launching a pair of landers. With the philosophy of 'those that never try always fail', the Soviets launched all four spacecraft in the summer of 1973, knowing they were most likely doomed.

Unsurprisingly, Mars 6 died first, just two months after launch. Ground Control continued to send the spacecraft commands, just in case, but there was no sign they were being heard. Mars 4 suffered a computer failure and couldn't correct its course, meaning it missed Mars by 2,000km. Mars 5 managed to reach orbit, but developed a leak in the pressurised instrument cabin and died three weeks later.

Mars 7, one of the landers, reached the planet. It was down one transmitter, but everything seemed to be going well when it released its lander on 9 March 1974 … until it fired its thrusters in the wrong direction, making it miss the planet entirely.

Thinking all was lost, the Soviets were delighted to hear a signal coming in three days after Mars 7's demise … from Mars 6's lander! The orbiter had been picking up the signals from Earth after all but had been unable to communicate back. The lander operated entirely as planned, coming down in the Mare Erythraeum, a large dark plain, but at a speed of 61m/s, much faster than it could survive.

Although no one had held much hope for the class of 1973, the tantalising hints of success were incredibly frustrating. 'Close, but no cigar' was becoming the unwanted mantra of the Soviet Union's Mars exploration.

However, the Soviet Union weren't the only players trying to get a spacecraft to Mars. On the other side of the world, the United States were preparing their own mission to the Red Planet – Viking.

12

THE ROCKY ROAD TO VIKING

The United States' route to Mars took a little longer than that of the Soviets, as early plans to visit the planet were sidelined by the Apollo missions. NASA kept an eye on the Soviets' progress, or more aptly the lack of it, and soon began jokingly referring to a 'Great Galactic Ghoul' lurking between Earth and Mars and devouring all spaceships sailing through its domain.

Space monsters aside, NASA did want to travel to Mars. Although Lowell's canals had fallen out of academic favour, the idea of life on Mars still held the public imagination. Astronomers had been watching the planet through telescopes for centuries, noting how patches of light and dark spread out over the surface with the changing seasons, perhaps caused by vegetation growing during spring, only to die back when autumn came around. There were even suggestions that Mars's red hue was due to a blanket of crimson foliage.

Just three months before the flyby, the National Academy of Sciences released a draft of their *Biology and the Exploration of Mars* report, stating, 'We believe it is entirely reasonable that Mars is inhabited with living organisms and that life independently originated there.'

NASA's attempts to reach the Red Planet met with much better luck than the Soviet probes, and on 15 July 1965 Mariner 4 flew past the planet, sending back images of the surface. The public keenly waited for what they hoped would be the first signs of life on another world;* hopes that were quickly dashed. The images were of low quality, but were detailed enough to reveal a deeply cratered, dry landscape with no atmosphere, not unlike the Moon's. No giant canals; no advanced alien life; no sweeping fields of red vegetation.

'Mars, it now appears, is a desolate world', read an editorial in the 30 July *New York Times*, provocatively titled 'Dead Planet'. 'Its surface bathed in deadly radiation from outer space, it has very little atmosphere and has probably never had large bodies of water such as those in which life developed on this planet.'

The find was a blow to the public's perception of Mars. The narrative of travelling to Mars had always been driven around hopes of finding life. Discovering that there was likely none made justifying the expense of exploration difficult. By the mid-1960s, the enormous cost of Apollo was garnering many critics. The space programme was gaining few results, while the Soviets beat the United States to almost every major milestone. With NASA spending so much money sending humans to one lifeless body, the Moon, what was the point of heading to the Red Planet if it was just another dead planet?

One of the planet's staunchest allies was American astronomer Carl Sagan. He would go on to become one of the leading voices in popularising astronomy. As well known for his love of polo-neck shirts as for his arrogance-free attitude, Sagan's vivid explanations of the universe captured the imagination. Although he created dozens of works about space, he is probably best remembered for his 1980 hit series, *Cosmos*.

At the time of Mariner 4, 30-year-old Sagan was still making a name for himself in the emerging discipline of astrobiology – the study of how life could grow on other worlds and what that life might look like. Sagan was fascinated by the origins of life, not just on other planets but on our own. Earth was once an inhospitable wasteland – if it changed, could other planets have done the same? And if not, why not?

---

*   The engineering team at NASA were just as keen to see the first image, albeit for a different reason. Mariner 3 had previously had issues with its tape memory storage. Rather than risk Mariner 4 suffering the same malady, the team printed the long stream of numbers representing the image onto strings of paper and hung them on the wall. The engineer in charge of the tape recorder, Richard Grumm, coloured in the image using red and brown pastels. Although the cameras might have been black and white, the first image of Mars was processed in colour.

One of his key questions concerned the fundamental ingredients to create life. While researchers could look at our own planet to work out the 'building blocks' of life on Earth – water; an energy source; key elements such as carbon, oxygen and nitrogen – there was no way to be sure that these were universal requirements for all life. If Sagan wanted to investigate that, then he needed to look for life on other planets.

Announcements of Mars being 'dead' irked Sagan. Mariner 4 had taken images of 1 per cent of the surface at 1km resolution – under those conditions Earth would look barren, too. Just because there weren't towering cityscapes and planet-wide red forests didn't mean Mars was barren of all life. There could be thriving colonies of bacteria-like organisms on the surface. Finding them, Sagan and his fellow astrobiologists argued, would be just as fascinating as uncovering an alien civilisation.

But there is a wide world of difference between a puddle of bacteria and the intelligent aliens that science fiction had promised. Mars was pronounced 'boring' and the planet faded from the public perception, although not from NASA's.

For James Webb, NASA's Administrator through the Apollo years, the Red Planet provided a beacon to lead the agency forward. He foresaw that once an astronaut first set foot on the Moon, NASA would change. The agency's budget was already being throttled by Congress. Without the thrill of the Space Race to drive funding, the agency could get stripped back to the bone.

The biggest fear for Webb was losing the Saturn V rocket. NASA had spent a decade developing these incredible machines, which stood 111m tall, weighed 2.8 million kg and could deliver up to 7.6 million lb of thrust (around 7,500 times that of a Boeing 747). But each one cost over $185 million (about $1 billion in today's money). If NASA had to tighten its belt, the Saturn V would be the first casualty.

Webb's solution was to plan missions requiring the Saturn V, giving Congress no option but to keep funding the building of them. One such mission was a lander called Voyager Mars.* Initially, this would use the smaller Saturn IB rocket, but Webb saw an opportunity for a new flag-ship mission.

---

*   At the time, it was mostly referred to as just Voyager, but as that name is now more commonly associated with the Voyager 1 and 2 missions that explored the outer planets in the 1970s and '80s, I'll refer to this Voyager as Voyager Mars.

This meant a serious upscaling of the project. The lander would now be packaged up with an orbiter to scout the way, letting the mission controllers pick the perfect landing spot. The orbiter would be based on Mariner 8 and 9, which were both scheduled to head towards Mars in the next few years, and the lander would bear a striking resemblance to the Surveyor lunar probes.

Unfortunately, the mission quickly became mired in its own ambition. As Voyager Mars grew bigger, so did its cost. The more expensive the mission was, the less acceptable failure became. To ensure success, redundant systems, checks and tests were bolted onto the mission and the costs began to spiral ever upwards.

Then came dissent from within the space community. Geologists and engineers thought a smaller series of missions would be best, slowly building up our knowledge of Mars over time. One such advocate was Bruce Murray, a planetary scientist from Caltech. Murray had worked closely with the twenty-two photographs sent back by Mariner 4, and saw that Mars was an interesting planet in its own right with a rich geological history just begging to be explored. There was more to Mars than its potential to host life.

Astrobiologists such as Sagan, however, knew small missions would never be able to deliver the answers to their questions about life on Mars. Practising with incremental missions was a waste of time when they had the technology to send a large-scale mission straight away. If landing the first man on the Moon would be a political coup, just imagine what being the first team to discover we weren't alone in the universe would do.

This central conflict between those wanting to explore the cosmos in small increments and those in favour of taking giant leaps runs throughout all planetary exploration to this day and nowhere is that more keenly felt than in relation to Mars.

Voyager Mars wasn't to be. Even before the Moon landing, political support for spaceflight was dwindling. On 21 February 1967, three astronauts were sealed inside an Apollo command and service module, preparing for an upcoming test flight. As the crew were working through their checklist, a fire broke out. Within moments the cockpit was in flames. The crew scrambled to get out but there was no handle on the inside of the hatch.

'We've got a bad fire – let's get out,' one of the astronauts cried over the intercom, his voice too garbled to make out who. 'We're burning up.'

Ground Control rushed to get the hatch open, but it was too late. All three men were dead.

The tragedy made people question the sense of the Apollo effort, and NASA's funding was squeezed even more.

The main fear of Congress was that Voyager Mars was a precursor to yet another expensive crewed project. Webb managed to keep his funding for a time by assuring the politicians it was a purely scientific endeavour, only to be undermined by his own people. In July 1967, the Manned Spacecraft Centre – without Webb's approval – sent out a general call for crewed mission ideas to Venus and Mars. Voyager Mars's opponents seized the call as proof that programme was a back door to a human mission and in October 1967, Congress killed the project.

This wasn't a vote against planetary exploration – just a plea for NASA to scale back their ambitions. NASA went back to the drawing board to create a cheap alternative to Voyager Mars. The first plan was to build a hard lander encased in a balsawood shell, but this was a swing too hard in the other direction towards the overly simplistic. They needed something in the middle.

The answer NASA arrived at was to use an orbiter–lander pair, much like Voyager Mars would have been, but with a vastly simplified scope. The agency would follow the Soviet philosophy of doubling up and send two of these spacecraft to Mars. If by some miracle both made it, they would maximise their usefulness by landing on opposite sides of the planet, where they would operate for a minimum of ninety days on the surface.

The missions were called Viking and would be built by JPL – which had essentially become NASA's planetary exploration wing – and ready to fly in 1975. However, the path from vague idea to finished spacecraft is never easy, and Viking would prove to be no exception.

Just like the Lunokhod and Surveyor missions on the Moon, the main problem was the tension between the various teams involved. The scientists wanted to build a fantastic science station. The engineers wanted the spacecraft to stay in one piece and working as long as possible. The administrators wanted everything to come in on time and on budget. The three rarely made good bedfellows.

Take Viking's cameras. Today, we have a vast array of digital cameras to choose from; not so the first Western lander to Mars. The cameras needed to be simple, lightweight and power efficient, with a low chance of failure.

What they came up with bore little resemblance to cameras as we know them today. Whereas modern sensors have millions of pixels, Viking had just twelve photodiodes (sensors that convert light into electrical energy, telling you how much light is shining on them). This single row of pixels rotated slowly, scanning across the scene to create an image. Each picture would take time, but NASA were pretty confident their camera would be fast enough to capture even the speediest of rocks as they sat motionless on the Martian surface. There would be two cameras, placed side by side to produce a 360° panorama of the landing site.

These details seemed to be the only things the different arms of the agency could agree on. The science team wanted a resolution of 0.04° – capable of making out something measuring 7mm across from 1m away. However, the engineers maintained this was excessive. To prove there was no scientific benefit to be gained, the engineering team compared a mock image with the high resolution that the science team demanded compared with one they deemed more feasible and asked people to tell them apart. Their grand scheme ultimately came undone when everyone identified the high-resolution image. The final camera resolution was set at 0.04°.

The engineers did have the last say on many other practical matters, though, such as size and weight – no amount of scientist wheedling would get the launch rocket to carry a heavier payload. One such victory for the engineers was over the camera placement. The science team wanted the cameras placed on a 2m-high mast, so that they had a clear shot to the horizon. But such a tall mast would be heavy, complex and fragile. After much wheeling and dealing, they ended up with the compromise of a design for a 1.5m mast, which somehow ended up being 1.3m on the finished rover, much to the annoyance of the science team.

There was a fourth arm of NASA making its dominion felt during the design process – planetary protection. During the Ranger and Surveyor days, the design teams had managed to avoid the most damaging sterilisation procedures owing to the fact the Moon was almost certainly barren. The central reason for going to Mars was that it might host life – there would be no casual waving off procedure this time.

Although sterilisation procedures had been refined since the probe-destroying days of Ranger, the Vikings would have to be scoured, washed and baked to 110°C for 40 hours. Any of the usual ways of storing the camera's images – magnetic strips or films – would be destroyed by the process. Instead, the camera would have to live stream its images directly to the orbiter, which would save the data and transmit it to Earth later.

Once the camera design had been agreed and built, the engineers were happy to let the cameras fly to Mars having never actually taken an image This plan so concerned the head of the Viking surface team, Thomas Mutch, that he heckled the engineers until they let him try out the camera in the nearby car park.*

But a lander is more than just a hunk of metal with a pair of cameras on top. Each of the landers was powered by a radio thermal generator (RTG). This is essentially a lump of radioactive material, in this case plutonium, which sits inside the lander, giving off heat that is turned into electrical power by a device called a thermocouple. This would power the lander for the required ninety days but could potentially last for several years.

They also carried transmitters to communicate with Earth. A meteorology package contained everything a young lander needed to keep an eye on the Martian weather – temperature sensor, wind-speed sensor, wind-direction gauge. A seismometer would keep an ear out for any marsquakes that might shake the ground. There was also a robotic arm that would allow the lander to take samples of the Martian soil. It was here where things really got interesting because these samples would be used to hunt for life.

Hopes of finding life on Mars had been reinvigorated by the flyby of Mariner 9 on 14 November 1971. Its images showed features on the landscape that looked remarkably like river-carved canyons, flood plains and all manner of features created by flowing liquid. Although now dry, water might once have flowed across Mars in the past. Could there still be enough locked up in the dirt to keep a microbe-like creature alive?

Viking would have not one but three different experiments to look for life. It was one of the reasons why the spacecraft would have to undergo such strict, memory-destroying, planetary-protection protocols – imagine the embarrassment of claiming to have found life on Mars, only for it to turn out later you'd actually just discovered germs someone had coughed up onto the lander.

---

* The cameras got a much grander outing in August 1974 when a flight spare made a trip to the Great Sand Dunes National Monument in the State of Colorado. As often happens when you give anyone a shiny new toy to play with, things soon took a turn for the frivolous when someone produced a pair of turtles and a snake. Unfortunately, both snake and turtles proved too speedy for the slow-scanning camera. The imaging team proved better behaved and stood still while they were being imaged ... then nipped around the back so they could reappear further down the group photograph – Thomas Mutch managed seven appearances!

A technician checks the soil sampler on the end of Viking's robotic arm. This device took a sample from the surface, then delivered it to the on-board labs to test for signs of life. (NASA: www.nasa.gov/image-feature/technician-checks-soil-sampler-on-viking-lander)

The three experiments all followed the same basic method – take a sample of Mars dirt, do something to it, see what happens. Specifically, they looked at what gases were given off that could indicate signs of life.

The first test was called the Gas Exchange experiment, headed up by Vance Oyama from NASA's Ames Research Center. The dirt was fed with nutrients, both with and without water, before measuring the levels of oxygen, carbon dioxide and methane given off.

Second up was the Labelled Release experiment, led by Gilbert Levin from Biospherics Inc. This again fed the sample with nutrients, but labelled these with a rare form of carbon, carbon-14. If the nutrients were metabolised by a lifeform, then this special carbon would be released. To make sure the nutrients weren't just being processed by some non-biological, chemical process, they'd repeat the experiment with some soil that had been superheated to sterilise it.

Lastly was the Pyrolytic Release or Carbon Assimilation experiment, led by Norman Horowitz from Caltech, which aimed to find life without either water or organic nutrients, since in theory any life on Mars should already have everything it needs to survive. The experiment flooded the chamber with carbon-14-tagged carbon monoxide and carbon dioxide, then waited to see if the carbon-14-tagged gas was released after being consumed and broken down by Martian bacteria.

Alongside these, Viking also carried the gas chromatograph mass spectrometer to give a better understanding of the base chemistry of Mars. This heated the soil until it vaporised, then passed the gas through a mass spectrometer – an instrument that could measure the weight of molecules and atoms. As each element has a specific, well-known mass, scientists used this information to work out what elements were contained in the sample. Although it wouldn't confirm what specific minerals and compounds were in the soil, the ratio of elements would give a good guide as to what might be present.

Organising the experiments should have been fairly straightforward, only there was one big problem – the three men in charge of each experiment hated each other. Rather than working harmoniously towards a unified scientific goal, they argued and fought to undercut each other's experiments. Horowitz, in particular, was critical of his fellows' work, publicly calling his colleagues 'irrelevant'. It was only through the work of the biology team lead, Howard Kline – nicknamed 'the Rabbi' for his heroic efforts in keeping the peace – that the three seemingly got through the project at all. This is just another example of how the biggest obstacles to landing on another planet are often the people trying to achieve it.

With all the infighting between geologists, engineers and astrobiologists, the costs of Viking were steadily rising. It seemed that the 'cheap' alternative to Voyager Mars would soon outpace the cost of the mission it was meant to replace. Plans for a back-up Viking 3 were scrapped to save money, despite being partially built. With no follow-on plans for Mars on NASA's horizon, if Viking failed it would be the end of US Martian exploration for some time.

Despite all this, the probe was built and by the summer of 1975, the spacecraft was ready to finally find out if there really was life on Mars.

13

THE VIEW FROM MARS

A few hours past midnight on 20 July 1976 at JPL. Despite the early time, the car park is already full. People dash from place to place in the dark. They have come for one purpose – to watch Viking 1's lander make its descent towards Mars.

At 1:51 a.m., the lander detaches from the orbiter and begins its long descent towards the Martian surface.

For many of the teams assembled, there is no real reason to be here. The true work for the teams controlling surface operations won't begin for several hours, once … if … the lander safely makes it to the surface. But the idea of coming to work later and missing this moment, the culmination of their work for the last few years, is unthinkable.

Work groups gather together in their assigned rooms, excitedly talking through the work to come. Occasionally, a lone engineer or scientist sits in a quiet room, preferring to pass the wait alone. As the clock ticks ever closer to zero hour, the landing team continue to check their spacecraft. In truth, there is little for them to do either. The landing is entirely automated, and the instructions were uploaded the day before. Viking 1 is so far away it takes its radio signals 19 minutes to reach the Earth, making operating it from Mission Control impossible.

At 5 a.m., all conversation stops. An overwhelming silence covers the entire facility, the only sound coming from the mission controllers reading out Viking 1's progress as it falls towards the surface at 4km/s.

'After years of waiting, hoping, guessing, the end rushes towards us – too fast to reflect, too fast to understand,' Thomas Mutch recalls in *The Martian Landscape*.

The spacecraft suddenly registers a huge G-force as it slams into the upper atmosphere. The worst of the shock is absorbed by the heat shield as it burns away.

At 5.9km above the surface now, the parachute has deployed, slowing the spacecraft even more. What's left of the heat shield drops away, and the lander deploys its legs.

After 51 seconds, the parachute cuts away. The landing thrusters kick in. The ground grows closer. Closer. Closer.

Then finally, at 5:12 a.m. the call comes through – 'We have touchdown.'

The entire building erupts in noise. People grab each other, shout and cheer. A Viking helmet, complete with apocryphal horns appears on the head of a young team member. But the celebrations are short. Now, it's time for the real work to start.

Although the landing of Viking 1 may have been textbook, its journey to get there was far from it. The probe launched on 20 August 1975, Viking 2 following along on 9 September. Viking 1 reached the Red Planet on 19 June 1976. The initial landing was scheduled for the US federal holiday on 4 July, Independence Day. Alas, the dual celebration of US supremacy was not to be.

The initial landing site had been chosen from Mariner 9 images as it looked like the smooth mouth of a large flood channel, the perfect place to study the history of water on Mars. But images from the Viking 1 Orbiter revealed it was filled with craters and islands carved out by the channels – far from an ideal location for landing a half a billion-dollar spacecraft.

After much arguing, Viking project manager James Martin Jr eventually ruled in favour of a flatter ancient riverbed in Chryse Planitia, 22° north of the equator. By chance, the new landing date would coincide with a different anniversary – 20 July, seven years after the Apollo 11 landing.

Viking 1 landed perfectly. Its squat form resembled a bug, albeit one as tall as a person. Its wide, flat body perched on three spindly legs with various booms and robot arms reaching out to taste the Martian environment, gaining humanity's first ever close-up view of Mars.

After landing, the camera popped up on its stalk, and within the hour its first images were making their way across the 549 million km from Mars to Earth. As it came in, line by line, the image showed the area around the lander's base, strewn with pebbles. There was an incredible level of detail, so sharp that you could count the rivets on the lander's foot that jutted into the image.

The next picture panned the camera up towards the horizon. The landscape was scattered with rocks of all shapes and sizes. It seemed the geologists on the team would not suffer from a lack of specimens to examine. But it was the sky that would captivate the imagination. Many had expected to see the blackness of space hanging over the Martian landscape. It turned out a bright sky arced over the planet, because dust in the air was scattering the light, just as our atmosphere does on Earth.

The next day, the first colour image came in, created by taking black and white images through coloured filters and overlaying them. It was time for the world to see just how red the Red Planet was.

Initially, the answer seemed to be not very. The first pictures released, showed a brown surface under a steely grey sky, but these images had been quickly thrown together by NASA to see what they were dealing with before the various layers were properly calibrated. Alas, NASA had failed to understand the media's voracious appetite for colour photos. Within 30 minutes of the pictures arriving on Earth, they were on international television with news anchors delighting in the familiar blue sky.

They were disappointed the next day when the Viking team released the correctly balanced photograph, revealing the sky was in fact pink. Some in the press corps even booed the image, asking if tomorrow the sky would be green. Carl Sagan was quick to call them out on their 'typical Earth-

The first ever successful image from Mars's surface taken by the Viking 1 lander just a few minutes after touch town. The left side of the image appears distorted as dust kicked up by the landing thrusters was still settling while the camera was scanning through this region. (NASA/JPL-Caltech: photojournal.jpl.nasa.gov/catalog/PIA00381)

chauvinist response'. Mars was an alien world after all. What was the point of going all that way if it was just the same as Earth?

After the initial excitement of the first landing, it was time to settle into a routine. With only a ninety-day expected lifespan, the probe began the most critical tests as soon as possible.

Almost immediately, the mission hit a roadblock. On the second Martian day, known as a Sol, the robotic arm jammed. The culprit was a locking pin that held the robot arm in place during transit and was now stuck.

Rather than Surveyor's scissor design, Viking's robot arm was much sturdier, using a metal tube that could extend out to 3m. The shoulder joint allowed the arm to swivel through 108°, left to right, while a hinge moved it up and down. The arm could place instruments on the surface, gather samples for the life experiments, scratch at the surface and even had a mirror to help the camera look around. It was key to almost every major Viking experiment – a stuck pin could ruin the mission.

After several days of wiggling its arm, Viking 1 managed to shake the pin loose, but the incident spooked both science and engineering teams. They were now loath to do anything without first getting a photograph showing the robot arm was correctly positioned, slowing operations.

As one sol is 24 hours and 37 minutes long, the days on Mars and the days on Earth slowly drift apart before lining up again. Crews worked through the night, piecing together images from another world as they came in. What a feeling that must have been – knowing you are the first human to ever lay eyes on it.

By August, Viking 1 was well under way. It was time for Viking 2 to take the spotlight as it arrived in orbit. After the resounding success of the first lander, you might think the atmosphere would be ebullient, but the opposite was true. After all the Soviets' troubles, the United States would be foolish to hope for both landings to go to plan.

Once again, the 'smooth' landing spot selected by Mariner 9 was anything but, so a new site was selected in Utopia Planitia, a little further north than Viking 1 and on the opposite side of the planet. On 3 December 1976, Viking 2's lander separated from its orbiter.

A moment later, the orbiter reported signs of distress.

What was wrong? Had the lander not disconnected properly and was now being dragged around Mars by an umbilical cord? Or was it at this moment hurtling towards the surface, out of control and doomed to crash?

Frantic to find out what the problem was, the team picked up the weak signal of the low-gain antenna, which was omnidirectional. Something had caused the orbiter to roll away from its lock on Earth, but the lander was still alive. More than that, they couldn't say.

The only thing to do was wait. So they waited. With no data to distract or reassure, dread began to rise that the 'cursed planet' might be about to claim its next victim. Over the loudspeaker, some unknown controller whispered 'Come on, baby', echoing the prayer of every person present. Just as the engineering team began to understand what failure felt like, the call came.

'We have touchdown.'

The room erupted, all the more joyous for the moments of panic that had come before and the case of champagne that appeared in the control room. They'd done it. They'd landed not once, but twice on the surface of Mars.

Viking 2's first panorama revealed a crooked horizon. The lander had come down on a small boulder, tilting it to an 8.2° angle. The angle wouldn't affect the mission, but the boulder could have pierced the lander's underbelly. If it had, then the delicate instruments inside would be exposed to the cold Martian night, conditions they hadn't been designed to survive. Anxiously, the team waited to see if Viking survived its first night. It did. All looked well for both spacecraft to make it through their ninety-day lifespan.

The landers were hard workers, taking hundreds of images of the Martian surface during their missions. And for landing sites that had been chosen to be bland and flat, they were surprisingly diverse.

Just 8m away from Lander 1's position was a worryingly large 2m-wide rock named Big Joe.* The lander could easily have come down on top of it and would not have survived. After Lander 2's close call, it was a timely reminder that while planning, calculating and testing were all important for landing on another planet, so was good luck.

To get a better understanding of the surface of Mars, Mutch's imaging team had to combine several images taken at different times. A single image gave no sense of depth, but by looking at the changes in shadow angle and length as the Sun moved across the sky, image analysts were able to gain some sense of scale and texture.

---

*   It was originally called Big Bertha, but a women's liberation group complained, and the name was changed.

Taking in the view, the landscape was littered with rocks, with fine sand-like dust filling the spaces between. While to the untrained eye this jumble might resemble a junk heap, to geologists it was the key to unlocking how Mars came to look the way it does.

Most of the big blocks of rock appeared to have been created by meteor impacts throwing boulders across the planet. The edges of craters were visible in some images but, unlike the Moon, they showed some signs of erosion, probably from the wind. There were a few rocks that looked like they might have been brought in by an ancient flood, a supposition backed up by the Viking orbiters finding evidence of flooding just 60km away from Lander 1's location.

While these were interesting, all these rocks had come from other locations. Without knowing exactly where they originated it was harder to put them into a wider context. Fortunately, Lander 1 hit geological pay dirt when it spotted large areas of dark-coloured rock. As these poked up from beneath the sand, rather than having been dropped on top of them, they were probably bedrock – rock that has stayed in the location it was created.

Viking 2 found so-called 'enigmatic troughs' – linear depressions 10–15cm deep. One theory is that they were created by the freezing and thawing of water locked up in the soil, as a similar process happens to the permafrost on Earth.

However, the images that had the most public appeal weren't ones of rocks or the pink Martian sky, but those where you could see the landers themselves. The contrast between the wild Martian landscape and the hard lines of the spacecraft captivated the imagination. It was all too easy to imagine a future where it would be astronauts taking those photos, snapping a selfie against the Martian backdrop.

While the images were giving a general overview of Mars, it quickly became apparent that each picture raised more questions than it solved. For instance, several of the rocks appeared 'sponge-like', pitted with tiny holes. Were they meteor impacts? Air bubbles found in volcanic rocks? Spots where a weaker mineral had once been but had now eroded away? It was impossible to say from pictures alone.

Even basic facts were argued over, such as how many different rock types were visible. The main problem was that on Earth, geologists don't simply look at a picture to determine the history of a landscape. They pick rocks up, feel their texture, study them under a microscope and, in some cases, give them a lick (although planetary-protection laws would no doubt prevent the latter, even if it were possible).* Instead, they would have to rely on the lander's labs to get a closer look.

The first job was to sniff the air, finding it to be mostly made of carbon dioxide, with traces of a few other gases, including argon, suggesting past volcanism as it had done on Venus. Next, it was time to give the ground a lick.** On opposite sides of the planet, the Vikings scooped up a sample of the soil, delivering it to the experiments that everyone had been waiting to hear from – the ones that would say whether or not there was life on Mars.

When the results first came in, there was a flurry of excitement – some of the results seemed to be positive. The Labelled Release experiment, which tagged nutrients with carbon-14, found that carbon was being released from the soil as if it was being metabolised.

Although NASA took pains to point out these were preliminary results, and incredibly uncertain, they made headlines around the world – which was unfortunate. The other two experiments showed no signs of life, as did a second round of the Labelled Release experiment.

The final nail in the coffin came from the mass spectrometer that analysed the bulk composition of the soil. There were no organic molecules. These are complex, carbon-containing chemicals. Although they can be created without life, they are the building blocks of biology, found in all Earth lifeforms. In their absence, the results of the other experiments were deemed moot. To this day, the results are classed as 'inconclusive', although Labelled Release leader Gilbert Levin still argues that his experiments found signs of life on Mars.

This doesn't mean that life doesn't exist, however. The arm could only dig 15cm into the topsoil. This upper layer has been irradiated by the Sun for millennia and was later found to contain hydrogen peroxide, a chemical we Earthlings use to kill bacteria. It's unlikely that any life, at least any that could be found by Viking's experiments, would be found living under these harsh conditions.

All four spacecraft, landers and orbiters operated far beyond their expected lifespan, solidly working away for sol after sol. Lander 1 lasted

---

*   I'm informed by geologist friends that this is by no means a universal habit. Rock picked up off the ground is generally not sanitary and sometimes even toxic. You don't want to pick up something thinking it's harmless only to end up with arsenic poisoning. So please, don't go around licking any rocks at home. Seriously.

**   No seriously. Don't.

until 13 November 1982, and Lander 2 managed 1,281 sols before its batteries failed on 11 April 1980. During that time, Viking gained an incredible view of the planet from above and below, watching Mars's changing seasons. In the winter months, the landers were surrounded with festive patches as frost grew in their shadows, although whether the frost was made of water, carbon dioxide or a mix of the two was unclear.

The mission had been hugely successful, but NASA let it down in one major way. Viking had been built up in the public consciousness as a quest to find life on Mars. That quest had failed.

Although the Vikings gave an incredible view of the Red Planet, the public ultimately held the purse strings. Their final verdict was that Mars contained no life and was therefore dull. In the 1980s, the United States moved its eyes away from planetary exploration. The Skylab initiative was building space stations, while a programme to build a fleet of reusable shuttles was in full swing. Although NASA proposed returning to Mars with a rover named Viking 3, the financial and political will for the $1 billion mission just wasn't there. The project was shelved. The prospect of the United States returning to Mars began to dim as America forgot about its love affair with the Red Planet.

14

THE RUSSIANS RETURN

It was now the Soviets' turn to make another run at Mars. Although their initial failures at Mars had been costly in every sense, the Venera missions had proved that the Soviets could routinely land on another planet – and a hostile one at that. Now in the early 1980s, and with the United States apparently losing interest in Mars, perhaps it was time for the 'Red' nation to return to the Red Planet and break the Curse of Mars.

The Soviets had learned that competing with the United States directly was a fruitless endeavour. Anything they could send to Mars's surface would be a pale imitation of Viking. Even if they sent a lander to another area of the planet, such as the frozen poles, they'd be compared to the NASA probes.

If they couldn't land on the planet, however, perhaps they could land on one of its moons. Mars has two tiny ones, Phobos and Deimos. The closest in of these, Phobos, is just 27km wide. It's so small it doesn't even have enough gravity to pull itself into a sphere. The lumpy space rock orbits a mere 9,000km from the surface, zipping around the planet once every 7 hours and 39 minutes. The first ever landing on another planet's moon was just the kind of landmark mission the Soviets wanted.

In terms of science, the moon's diminutive size means it hasn't undergone many of the large-scale geological processes that shape the planets, leaving it largely unchanged since its formation. Studying the moon would be a chance to see how the ravages of the solar wind, cosmic rays and meteorites affected celestial bodies caught in this early stage of their formation.

The project was officially announced on 14 November 1984, initially aiming for the 1986 launch window, although this quickly slipped to 1988.

As usual, there would be two spacecraft. In an effort to break the bad luck of the previous missions, they would not be named Mars. Instead, they would be Phobos 1 and 2.* As the moon was too small to orbit easily, the main probe would instead fly around Mars making frequent sweeps past the moon. After a thorough examination, it would drop its lander onto Phobos.

Landing on the moon was no trivial matter. Phobos's diminutive size meant the spacecraft wouldn't have to fight against gravity, but size is also related to how fast you need to be going to achieve escape velocity and get away from a planet's surface. On something as small as Phobos, the rebound from hitting the surface could be enough to launch the lander into space again.

As the Soviets were unsure of the best approach to make sure they stuck the landing, there would be two landers on each mission. The first was the Long-term Automated Lander (LAL), which would fire harpoons into the rock to anchor itself. The second was a 'hopper', shaped like a ball flattened on one side, that would closely match Phobos's speed before release. After coasting in at a low speed, it would then drop to the surface, rounded side down, before rolling onto its flat side. Then an internal spring would take advantage of the low gravity, allowing the lander to hop around to as many as ten different locations.

As with the Vega flights, which were making their way to Venus around the same time, the level of international co-operation was very high. Aleksandr Zakharov, Phobos's project scientist, said:

> We hope that this effort and the increased sharing of scientific data by the Soviet Union and the United States will result in researchers of all nations no longer being confined by oceans and borders in the pursuit of scientific knowledge.**

The project was incredibly complex, and although he lauded it for its ambition, Carl Sagan referred to Phobos as 'hair-raising'.

Phobos 1 launched from Baikonur on 7 July 1988. At that time, the Soviet Union, led by Mikhail Gorbachev, was attempting to be more

---

*   Seeing as Phobos comes from the Latin for fear, I'm not entirely certain their nominative hopes for fortune were well founded.

**   A.V. Zakharov, 'Close Encounters with Phobos', *Sky and Telescope*, July 1988, pp.17–18.

transparent in its affairs, a policy known as *glasnost*. As such, the launch was witnessed by journalists and scientists from around the world. Fortunately, everything went well at this stage, as did Phobos 2's launch five days later.

But changing the programme's name from Mars to Phobos wasn't enough to save the missions from the Red Planet's curse. On 2 September, a computer command was sent to Phobos 1 to turn on the gamma ray spectrometer. A single hyphen had been left out of the code, transforming it into an order for Phobos 1 to shut down. There was no way to turn it back on.

It was a simple mistake, a single digit typed incorrectly, but it was a mistake that was symptomatic of the discord within the Soviet space pro-gramme. They still didn't have a central governing body like NASA, and the various disparate agencies frequently didn't communicate well. In this case, while Moscow oversaw the mission, Ground Control was based at the long-range tracking centre in Yevpatoria, in the Crimea. They were the ones responsible for checking the code, but the necessary equipment at Yevpatoria was broken. Moscow were unaware the code hadn't been checked and blindly uploaded the commands.*

With only one probe left, the team took a lot more care over Phobos 2 when it arrived at Mars in October 1988. However, despite their efforts, the outlook wasn't hopeful. There had been multiple different failures, all apparently isolated from each other.

The problem turned out to be a fault in how the spacecraft's three redundant computers operated. To protect against a faulty computer, the three would 'vote', meaning two healthy processors could outvote one erroneous one. No one had thought about what would happen if two of the processors were unreliable, which was exactly the case on Phobos 2. One was broken outright, while the other was proving troublesome. The third couldn't outvote the two dead processors.

There was still a chance they could land on the tiny moon, if the Soviets could speed up the timeline before the ailing spacecraft gave up entirely. Phobos 2 performed its first pass of the moon on 1 February, at a distance of 864km. It passed by again on 21 February, managing to snap a few pictures.

---

*   Because they didn't want to stress the Phobos 2 team, no immediate disciplinary action was taken, as management was apparently taking the advice of Lavrentiy Beria, Lenin's Secret Service chief, to heart – make them work for now; we can still shoot them all later.

By 3 March, the science team had determined the accuracy of Phobos's orbit to within 5km and set the landing date for 9 April. They would never get the chance. During another imaging run, the spacecraft cut out. Although Ground Control managed to get the signal back for 13 minutes it was, in the words of Roald Sagdeev, the Director of the Institute for Space Research, 'a very weak, unintelligible signal indicating that the spacecraft was in an uncontrolled state. It was the last message from the dying Phobos 2.'

It was officially declared dead on 16 April 1989.

The failures came as a shock after the unbroken successes of the late Venera missions. Although recriminations flew, the cause of the failure had ultimately been poor design and management.

There were plans to launch a back-up craft, Phobos 3, in 1992. However, the Soviet Union was having financial problems and instead sold the lander to the West to raise funds. Eventually, the government agreed to allow another run of Mars missions, this time with significant input from several European nations and the United States.

Initially, these were planned to have a Vega-like balloon, this time built by the French. The mission, Mars 94, would launch, surprisingly enough, in 1994. Two years later, a rover would follow. Two years after that, a sample return.

As work began in 1990, problems began to befall the project, not from within, but from the very top of Soviet society. The Soviet Union was running out of money and while the funds for the Mars project were promised, there was no guarantee they would materialise. As Mars 94 began to take shape, Soviet leader Gorbachev was trying to dodge *coup d'états* and hold his struggling nation together. He failed. In December 1991, the Soviet Union collapsed, dissolving to form Russia and the Commonwealth of Independent States.

Various arms of the space programme were now based in different countries, each of which were clamouring to set up something resembling a stable government. The lack of central command had now become a much bigger issue.

Amongst the turmoil, NASA took a step back from the project, while Germany and France, who were already building parts of the mission, increased their grants. Some funding from the Russian Government did filter through to the Mars 94 team but it came in intermittent bursts. Progress on the spacecraft stopped and started as the funds arrived and then ran out. And so Mars 94 limped on.

Despite the turmoil at the time, the new Russian nation found the time to finally set up a unified space agency in February 1992, called Roscosmos. One of its first decisions was to delay its Martian plans rather than risk launching a substandard spacecraft and Mars 94 became Mars 96.

Yet the nation's wider problems continued to impinge on the project. Power cuts resulted in the technologically advanced spacecraft being built by candlelight, while kerosene heaters kept the workers – who were often going unpaid – from freezing.

Eventually, the spacecraft made its way to the pad, but it was clear that this mission was the end of Russia's planetary dreams for the foreseeable future.

The mission was an orbiter carrying two golf-tee-shaped penetrating landers weighing 65kg each. They would hit the ground at a relatively high speed, burrowing themselves into the surface. Along their length, sensors and cameras would take in the surroundings. Additionally, it carried two stationary landers akin to a stripped-back Viking. Most excitingly, all four stations carried seismometers. By comparing signals across the planet, the quartet would create a network that could not only listen for marsquakes but work out where they originated.

Alas, the mission could not overcome the problems of its birth. It launched on 16 November 1996 but failed to leave Earth orbit. It crashed back to Earth a day later, the wreckage coming down somewhere in the Andes mountains of Chile.* Russia had hoped to break the Curse of Mars, only to fall deeper under its spell, ending their planetary dreams with both a whimper and a bang.

---

*   This was somewhat problematic, as it contained a plutonium power source. The Chilean Government was not happy that no one had warned them about potentially hazardous material crashing down on their heads. They were even less happy that no one came to get it – the Russians couldn't afford to. As far as anyone knows, there's a lump of plutonium out there to this day.

15

THE PATH TO PATHFINDER

In the wake of Viking, the United States' plans to return to Mars stagnated. Although some bandied around ideas of returning to the planet, public support just wasn't there. Mars was seen as a white elephant, talked about quietly in back rooms but rarely openly. The genesis of change came from an unexpected place – a handful of students at the University of Colorado, Boulder.

Steve Welch was never going to be one of the cool kids. During his childhood, the Florida teen would spend every night escaping his earthly troubles by looking to the stars above. In his early teens, he met Penny Boston. The daughter of theatrical parents, her youth had been spent riding elephants in the circus and being sawn in half by magicians. But she, too, had dreamed what life would be like on other worlds. The pair began dating and eventually married.

In 1976 both Welch and Boston were at Boulder, studying physics and biology respectively, when they watched the Viking mission touchdown. Boston later described it as 'a life-changing moment. Like watching Armstrong and Aldrin walk on the moon … It hit us all between the eyes that it was a planet. It was a place … it was an immediate human connection to the landscape.'*

They quickly became obsessed with the mission, and soon other Mars enthusiasts were drawn towards them. Boston shared an office with Carol Stoker and Chis McKay. Stoker's connection to the Red Planet was almost

---

*   Andrew Chaikin, *A Passion for Mars* (Harry N. Abrams, 2008), p.127.

spiritual, as her childhood dreams had been filled with images of an alien landscape covered in fields of tall red grass. Always out of place in her Utah hometown, she found kindred spirits in Boston and Welch. Meanwhile, McKay trawled through the Viking data, sharing his finds with the rest of the group. Finally, there was Stoker's boyfriend, Tom Meyer. Although his previous work had been building robots to explore the ocean floor, he now turned his expertise to thinking about how mechanical explorers could be used on Mars.

They were an oddball group, full of youthful confidence and wonder, tempered by keen scientific minds – the kind of people who sat around discussing hard science theory while wearing antenna-like deely boppers left over from a Halloween party. While the rest of the world forgot the Red Planet, they discussed ideas of how to terraform Mars, adapting the planet to be hospitable to life.

The idea had been postulated before – Carl Sagan himself had written a paper suggesting Mars's temperature could be raised by painting the poles black. The group discussed these at length, while Boston performed experiments in 'Mars jars'. These were vessels filled with a Martian-like atmosphere where she grew plants and nitrogen-fixing bacteria – a vital part of Earth's ecosystem that could be used to transform the Martian soil. They called themselves the Mars Study Project, although often went by the catchier moniker 'the Mars Underground'.

The group's work attracted the attention of Ben Clark, a former Viking team member who wrote a paper entitled 'The Case for Man on Mars'. He acted as an ambassador between the students and his former colleagues at NASA. Soon, 'closet Martians' from all areas of space research began making themselves known to each other.

But the Underground knew they weren't going anywhere if they remained in the closet, fearing being seen as 'crazy' for talking about returning to Mars. What they needed was to get everyone in the same room. It was space journalist Leonard David who initially came up with the idea of a Mars conference. With no one else willing to run one, perhaps this enthusiastic bunch of students should do the honours.

In April 1981, the group organised the first 'Case for Mars' conference (choosing to drop the 'Man' part from the name of Clark's paper to avoid potential sexism debates). They sent out an open invitation all over the world, asking everyone and anyone who wanted to come talk about the Red Planet to attend, hoping that at least some people would turn up.

Over 100 Mars enthusiasts crept out of the shadows: NASA exobiologists, who had escaped the organisation's regular budget cuts; academics who felt their work at Mars was far from done; the conspiracy theorists who wanted to find proof that a face-like feature seen by the Viking orbiters was built by an alien civilisation. They talked of the logistics of getting humans to another planet, the habitats they would need and the life-support systems to keep them alive. Most importantly, they talked about Mars, openly and freely, for the first time in years. 'The first conference was magic,' said Stoker. 'People walked out of there feeling like they'd been freed from prison … we broke the taboo.'*

Over the years, follow-on conferences attracted big names like Carl Sagan, Buzz Aldrin and Tom Paine, who had led NASA during the Apollo 11 landings. The last 'Case for Mars' was held in 1996. By that time, NASA was firmly supporting Mars once again, and sending missions back to the planet.

If there was one thing Viking had taught NASA, it was just how critically important the public was to their mission. Bruce Murray, now director of JPL – essentially NASA's planetary exploration wing – decreed that all missions should be 'purple pigeon' projects.** As well as a high scientific value, the missions needed to have a level of gravitas and excitement that the public could get behind, as it was ultimately the taxpayer to whom NASA was answerable. Public opinion had killed Mars exploration after Viking, but it was Viking that had lit the spark of enthusiasm in the next generation who had brought Mars back to the table.

Whatever mission came next would need to have something impressive to sell to the public. While a human landing would certainly have that, this was no Apollo situation. The astronomical cost would have many opponents. It would be better to start with robotic missions, with a long-term goal of one day sending humans.

Talking about hunting for life was a dangerous prospect. Even if – and that's a very big *if* – there is, or ever was, life on Mars, there were serious doubts about our ability to find it. Instead, the teams at NASA decided

---

*   W. Henry Lambright, *Why Mars: NASA and the Politics of Space Exploration* (Johns Hopkins University Press, 2014) p.81.

**   The term 'purple pigeon' arose as an antithesis to 'grey mice' missions, which were unexciting and timid. Why the pigeon was chosen as the symbol of excitement and daring, I don't know.

to rephrase the exploration of Mars not as a biological investigation but a geological one.

Mars, the planet, is itself a very interesting creature. It was once a volcanically active world, home to the biggest known volcano in the solar system, Olympus Mons. With a light gravity and no atmosphere pushing down on the rock, the mountain grew to an enormous height of 22km.*

The planet was caught in a peculiar part of its history. It was large enough to differentiate, meaning that when it was forming, the pressure and heat was high enough to melt the internal rocks. These then separated out with the heavy metals sinking to form an iron core, while the lighter elements floated to the surface to create a rocky crust, with various other layers in between. However, Mars was too small to stay warm. On Earth, the radioactive elements within the rock keep our planet's interior nice and toasty and, most importantly, fluid. With liquid, or semi-liquid, layers under the surface, our crust can grow and change via plate tectonics. Mars froze early on in its history, remaining trapped in its infant form forever, giving us a snapshot of what a baby rocky planet looks like.

With Mars back on the agenda, the question was what to send. The ultimate goal was to return samples to Earth. While geology was the way to sell the mission, the hunt for life was still in everyone's thoughts. Feasibly, the only way we'd find life on Mars was by allowing a human to study Martian samples in a full-blown laboratory. Sending a human and lab to Mars was still many decades away – technologically, logistically and politically. So, the only option was to return samples to Earth robotically.

Ideally, if you're gathering samples on another world you don't want to only pick up rocks where you happen to touch down. You want to be able to see a variety of things and track down the most interesting places. There were two ways of achieving this. You send many small missions, dropped at different locations across the planet. Alternatively, you can get mobile. Over the years, there were several plans to move around Mars, including a balloon that would float through the air during the day, then sink to

---

*   To put that in perspective, most commercial planes fly at 10–12km. It's a big 'un.

the surface when the night's chill shrunk the air inside the bag, and an aeroplane-like probe that could fly 600km.

In the 1980s, the planetary exploration team at NASA began planning a Mars rover. The first task was working out how to build one. Although the United States had learned much about sending a vehicle to another world from the Apollo's lunar rovers, these had all had an astronaut at the wheel. A Martian rover was different. Even when Mars is closest to Earth, it takes 8 minutes for a signal to make the round trip. At its furthest, that goes up to 48 minutes. If it was operated from Earth, the rover would have to take a picture, transmit it to Earth, wait for it to be evaluated, wait for the commands to be sent back, drive a few inches, then stop and repeat the process. It would be infuriating, like taking a hike where you had to wait 48 minutes between steps. Instead, they would have to have a rover that was semi-autonomous, capable of doing at least some of its own thinking.

Then there was the issue of getting the rover to Mars. It was much further than any rover had been sent before, around 500 million km compared to the 385,000km hop to the Moon. That requires more fuel, meaning a much lower weight allowance for the payload – and rovers tend to be heavy.

Last, but by no means least, with a sample-collecting mission is the problem of contamination. While contaminating a scientific reading with life from Earth is one thing, returning it risks contaminating our own planet, another issue entirely. If our own history hadn't shown us the genocide that results from bringing foreign microbes to new lands, then science fiction certainly has, with books such as the *Andromeda Strain* and films like *Life* showing exactly what could happen if we brought back a 'bad bug' from Mars.

With all these things to consider, NASA originally slated a sample-return mission called the Mars Rover Sample Return (MRSR) for the 1990s with the goal of bringing back at least 4.5kg of Mars rock. The study team was led by Donna Shirley. As an engineer at NASA since the 1960s, she had already worked on missions to both Mars and Venus. She was one of the few women to have made it to the upper echelons of JPL, and as such Shirley often found herself the only female voice in the room. Still, she managed to corral the youthful engineers of her team, to explore all the options first rather than rapidly focusing in on a single design.

The first decision was how the rover was going to move. On tank treads? Wheels? One working group at Carnegie Melon University suggested a complicated system of legs. While the idea of a mechanical spider scuttling around Mars was intriguing, this last idea was unfortunately too complicated and expensive, and quickly fell by the wayside.

After several efforts, it was mechanical engineer Don Bickler who hit upon the winner – the rocker-bogie. There would be three wheels on each side of the rover. The rear two of these would be linked by a U-shaped bar, which would then be attached by a second U-shaped bar to the fore wheel. It was an elegant design and could happily trundle over most obstacles. Despite the potential of the design, he had to fight for the funding to build a prototype and built most of it out of wood in his own garage from schematics drawn in felt-tip pen. It was worth it, though, and the design was chosen for not just the MRSR, but every rover since.

Another issue the study addressed was the question of autonomy. The idea of a spacecraft thinking for itself had only recently become possible. At the beginning of the 1960s, the computers NASA used to perform their calculations filled rooms and weighed tonnes. Stopping the lander from damaging a motor by extending an arm too far or driving itself of a cliff was largely down to the human operators knowing their machine. Sometimes, protective measures were crafted into the spacecraft hardware, but this made them impossible to override, as was sometimes useful.

By the time of the study, technology was improving to the point where a fairly advanced (for the day) computer could fit onto a rover. Using the input of cameras and other sensors, the rover would be able to sense the lay of the land, assess it for hazards, plan a path based on where the ground team ordered it to go and then execute it. Once it had reached the end of that plan it would stop and repeat the process.

The software team wrote a trial version of the software to do just that. It was slow but it worked and would be a darn sight faster than waiting 48 minutes for a command from Earth.

The team worked through every challenge a potential rover might face, evaluating every aspect of the mission. The study showed that NASA absolutely could send a sample-collecting robot to Mars. It also showed it would come with a hefty $10 billion price tag – probably more. It was just too expensive, and the project was shelved but not forgotten.

Before long, NASA was developing a new Mars mission, the Mars Environmental Survey, or MESUR. The aim was to launch sixteen to twenty landers across the surface of Mars. Together, they'd form a network analysing the planet's climate, terrain and seismic activity, creating a global picture of the planet from the ground.

While the entire project would cost over $1 billion, the individual landers would be relatively cheap. If one or two failed, it was no big loss and they could launch several on one rocket.

Rather than jumping in with all sixteen feet first, NASA decided to send a single test spacecraft – MESUR Pathfinder. The mission would be one of the first to be funded by the Discovery Programme. This was a new initiative set up in 1990 to supply low-cost, fast-paced missions. In summer 1991, MESUR Pathfinder was given a $150 million budget, handed to JPL to build and given a launch date of 1994. Mars had returned to its place in the planetary spotlight and now NASA was going back.

The mission came at a time of great change at NASA. In 1992, Dan Goldin took over as NASA's Administrator. He was a man with vision and intensity, as well as a soft spot for Mars. At the time he took the position, NASA's finances were becoming increasingly wayward. It was time to bring them into line and he did so with a new management philosophy – 'faster, better, cheaper'.

At first glance, this might sound like flawed logic. In most project management, its generally established that out of fast, good and cheap you can have, at most, two of those options. Push for all three and you end up with none.

But NASA missions were suffering from a serious bloating problem. Take the Mars Observer, NASA's first mission back to Mars since Viking, which launched on 25 September 1992. Everyone wanted their experiment on the orbiter because who knew when the next one would fly? The project's expense spiralled as increasing numbers of tests and checks were added to ensure it wouldn't fail.

Goldin argued it would be much better to have smaller but more numerous missions exactly like those being attempted with the Discovery-class missions and Pathfinder. Each would have a narrower scientific scope, but there would be so many more being flown that the total achievement per dollar spent would go up. And if one mission failed then, yes, it would be unfortunate, but not as catastrophic as the loss of something as big as the Mars Observer, which lost contact with Earth just as it was coming into Mars orbit.

While the press lamented the loss of a 'Billion-Dollar Satellite'* as a waste of taxpayers' money, Goldin seized on the moment to push through his changes and apply the ethos of 'faster, better, cheaper' across the agency. Pathfinder would be his standard bearer.

---

* The 'Billion-Dollar Satellite' is something of a misnomer. The actual cost of building the spacecraft was around $500 million. When factors like operational costs over the time the spacecraft would have been at Mars were added in it did it rise to a maximum of $980 million. But it served Goldin's purposes to emphasise the cost, making Mars Observer probably the only spacecraft to have had its cost inflated by NASA, rather than downplayed.

16

MARS IN A BOX

On 7 August 1996, Bill Clinton stepped up to a podium on the White House's southern lawn. That morning, NASA had released an extraordinary announcement. Geologist David McKay had found chainlike structures that could be the fossilised remains of bacteria inside a Martian meteorite recovered from Antarctica, named Allan Hills 84001. It was an extremely preliminary discovery, something both the scientists and Clinton hastened to emphasise. 'Like all discoveries, this one will and should continue to be reviewed, examined and scrutinised,' Clinton said in his statement on the discovery.

There were many things about the 'fossils' that didn't match up. They were much smaller than Earth bacteria, and the rock could easily have been contaminated in the 13,000 years it had spent lying on the ground after falling to Earth. But it was the latter part of Clinton's statement that resonated with the public:

> Today, rock 84001 speaks to us across all those billions of years and millions of miles. It speaks of the possibility of life. If this discovery is confirmed, it will surely be one of the most stunning insights into our universe that science has ever uncovered.

Media across the world clamoured to know what was being done to follow up on the discovery. Their eyes all turned towards a Martian surface mission that was due to launch in just a few short months – Pathfinder.

The team would soon be bombarded with questions about how the Pathfinder mission could be adapted to hunt out more signs of life, bring a sample of rock home and give answers everyone hungered for. The response was … it couldn't. The spacecraft was already finished and most of the way through testing. It was too late. Any answers the spacecraft could deliver would have to be with the equipment that had been decided on four years before.

Back in 1992, the MESUR Pathfinder mission was just coming together. The main goal of the lander was to test the technology that would be needed for the rest of the planned MESUR landers, but NASA weren't about to waste the opportunity to send scientific instruments to Mars. There was plenty of room for experiments to study the geology and weather of the Red Planet.

The lander would need cameras. Aside from the engineering benefits of being able to examine Pathfinder after landing, the agency was well aware of how important its pictures would be for PR purposes. Several years before, NASA Associate Administrator Burt Edelson had forced the Mars Observer team to go back to the drawing board when he realised the orbiter didn't have one on it, telling them, 'I'm not going to approve of any mission to Mars, or any other planet, that doesn't have a camera aboard.'*

Pathfinder's camera was called the Imager for Mars Pathfinder. It was a stereo imager with two lenses side by side. There would also be a meteorological station, known as the Atmospheric Structure Instrument/ Meteorology Package. This would regularly measure the temperature, pressure and wind speed and direction for as long as Pathfinder operated.

Another critical instrument was the Alpha Proton X-ray Spectrometer (APXS). Having been derived from an instrument developed for the Soviet Mars 96 mission, the APXS was similar to the Alpha–Scattering surface experiment carried on the Surveyor probes. However, rather than just using alpha particles to detect a rock's composition, APXS used protons and X-ray particles to detect a wide range of elements and molecules.

This was vital for understanding Mars – and would be a recurring instrument on planetary landers for years afterwards – but strapping it to

---

*   This philosophy has stuck at NASA. The 2016 Jupiter Orbiter, Juno, has only one visible light camera, JunoCam, an instrument whose sole purpose is taking images for public use. The fantastic images that were produced have kept the spacecraft in the public eye for years.

a stationary lander, even on the end of a robot arm, felt inadequate. Early plans for the landing system showed there was enough room to carry a micro rover. Finally, NASA could test out the ideas put forward during the MRSR study. They could mount the APXS on the rover, greatly increasing its usefulness, and the tiny rover would help to rally the public.

After convincing NASA officials that not all rovers needed to cost $10 billion, one was added to Pathfinder. However, the political manoeuvring to get it on the mission meant it wasn't funded by the same branch of the agency as the main lander. Instead, a section known as Code R, which controlled automation and robotics research, was footing the $25 million bill and their main aim was to test rover technology. Any scientific work by the rover would be secondary.

Once again, Donna Shirley was in charge of managing the rover team. Almost immediately, every science team began to hound her, claiming the camera needed to be placed just so or had to carry this specific instrument. She quickly laid down the law with what became known as Shirley's Rule: 'a requirement is not a requirement until someone pays for it'.

In many respects, the micro rover was just another one of Pathfinder's instruments. But the unique funding situation meant the two spacecraft were developed almost independently, which only added to the usual conflict between teams trying to create any space mission.

With lander and rover decided, the Pathfinder team began to work on getting them to the surface. That required some novel design work. The lander would be protected during its descent inside a box shaped like a tetrahedron – a pyramid where every side (known as a petal) is an equilateral triangle. If the lander came down on its side rather than on the base, the petal would open out to right itself before opening the other two petals, each carrying a solar panel to power the lander. Unfolded, Pathfinder would measure 3m wide. Once settled, several long masts would extend upwards, such as that carrying the camera and the communications antennae that would communicate with the orbiter.

The big question was how to get the lander down onto the surface. Thanks to Viking, NASA knew how to get through most of the atmosphere via a combination of atmospheric braking and parachutes. The problem was the last few hundred metres to the ground. Viking had used retrorockets, but these would contaminate the surface – Pathfinder needed a method that didn't use chemicals. The solution was one used by delivery companies for decades – bubble wrap. Rather than a plastic sheet, Pathfinder would be wrapped in twenty-four interconnected airbags, each

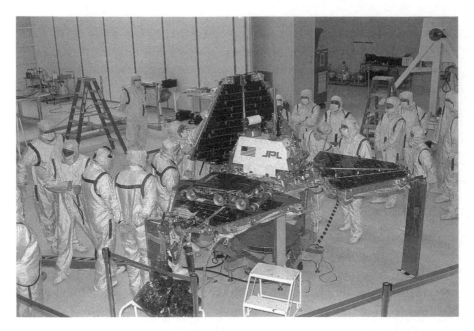

Engineers folding up Pathfinder's petals ahead of launch. The Sojourner rover can be seen attached to the front left petal. (NASA: www.nasa.gov/content/making-final-preparations-for-the-path-to-the-red-planet)

5m in diameter, which would inflate 8 seconds before impact to absorb the shock.

The airbags would need to be rugged. The lander could bounce as high as a ten-storey building and then land on a jagged rock. The material took inspiration from the Apollo moonwalking suits. It was made of several layers, the outer of which could tear a little, preventing the inner layers from popping. Once the lander had come to rest, the bags would then retract so as not to get in the way.

To sew them, NASA turned to the extremely talented seamstresses at aerospace manufacturers ILC Dover on the east coast of the United States in Delaware. Their qualifications were undeniable – one of their number, Eleanor Foraker, had sewn Armstrong and Aldrin's moonsuits, three decades earlier, the very suits Pathfinder's design was now cribbing from.*

---

*   Not nearly enough credit is given to the seamstresses at NASA, who are almost exclusively women. Their handiwork is found on nearly every mission, from the perfectly fitting thermal blankets around satellites to the intricately sewn fingers of an astronaut's glove. The precision of their sewing is the rival of any mechanical engineer's soldering.

With Goldin's 'faster, better, cheaper' ethos firmly in effect, both teams knew the days of ballooning costs were over – their budget was their budget. Chief Rover Engineer Bill Layman kept his team on task by finding the 'optimum level of stress'. Trusting his team's intuition, he pushed them to take leaps in the design and manufacture, such as going straight in with an assembly technique on the flight hardware without trialling it first. Layman said:

> There's an optimum point where they [those building the rover] succeed with ninety percent of their leaps. For the other ten percent they don't there's time to remake that leap or go a more meticulous route to the solution of that particular problem.*

One of the main ways the Pathfinder team saved money, particularly on the rover, was by using existing technology wherever possible. This was highlighted in the quest to find an appropriate central processing unit (CPU) – the rover's brain. As the rover was tiny, it could only have a small solar panel. It also couldn't have a large, heavy rechargeable battery. Instead, the battery would be pre-charged, and its use reserved for night-time operations, while the solar panel would have to be able to run all systems during the day.

The faster a CPU runs, the more power it draws. The CPU on the rover had just 0.75W to work with. In a typical desktop computer, the CPU's fan alone draws around 2W. The rover could go without a CPU entirely, relying on the lander, with its enormous solar panels and rechargeable battery, to do all the thinking. However, this quickly became unfeasible. The rover needed its own brain.

The solution lay not in a novel approach but an old one. The team uncovered a 20-year-old processor, the 80C85, which would be able to withstand the radiation and temperatures on Mars and, most importantly, was slower than a tortoise and required a tiny amount of power. This meant that everything else on board would have to operate just as slowly, creating a new set of problems.

For instance, the camera was a charge-coupled device (CCD). These have grids of pixel sensors, each of which record the amount of light falling on them as a charge. The brighter the light, the larger the charge.

---

*   A. Mishkin, *Sojourner: An Insider's View of the Mars Pathfinder Mission* (Berkley Books, 2003) p.109.

Normally, the sensor would be read off, line by line, in a fraction of a second. The 80C85 would take around 53 seconds. In operational terms, this was fine – it would take hours to send the image back to Earth, analyse it, decide what to do about it and then send the commands back to Mars. The problem was that as the sensor was sitting being read out, it would heat up, potentially changing the charge of the pixels and fuzzing up the image.

Fortunately, Mars is cold enough that the noise after a minute should still be low enough to get a decent image.* The limits did mean there would only be one processor, though. If it failed, the mission was over. According to Pathfinder's mission goals, the rover just needed to survive seven sols. In truth, most people were aiming for more like thirty sols and crafted their spacecraft to last as long as possible while staying in budget.

Not all corner-cutting went smoothly, however. One area of intense debate was how the rover would talk to the lander. The rover was far too small to communicate directly with Earth, so would relay its data through the lander's beefy transmission equipment. The leading candidate was a glorified walkie-talkie radio between the two.** Normally, these radios would have been specifically designed for the task, but NASA used off-the-shelf models, the Motorola RNET 9600 radio modem.

The vibrating crystals that controlled their transmission frequency weren't thermally controlled, so as the radio's temperature fluctuated, the frequency of the transmission would too. If it drifted too far, then the transmitter and receiver could end up talking at completely different frequencies. But there wasn't time or budget to replace the crystals. The issue wasn't a deal breaker – the team could get the pair to hunt across various frequencies until they found the signal – but it could have been avoided entirely with just a little more time and money.

The final rover was a tiny thing, just 63cm long and weighing only 10kg. It was covered in gold foil to help regulate its temperature, with a shiny blue solar panel on top and struts that wouldn't look out of place in

---

*　　To prove this, engineer Brian Wilcox shoved a test camera into an ice box. He printed the image out and hung it on the wall so that anytime someone said the camera wouldn't work, he could point to it and show the images were just fine.

**　　There was talk of using a thin cable, known as a tether. Pathfinder project manager Tony Spear was evangelical about these (his aversion to radios is believed to stem from a traumatic experience with a cordless phone), but he changed his tune after hearing about the robotic exploration of active volcano Mount Erebus in Antarctica. The tether snapped a few minutes in, ending the mission irrevocably.

a Meccano set. The rover looked like little more than a toy. And yet, the work it did on Mars would help to decide how every planetary rover after it would be built.

Such an important rover needed a fitting name. Sensing the potential for a PR coup, Shirley set about organising a public competition to name the rover. She decided their little explorer was female (despite the grumbling of the predominantly male design team) and asked students from around the world to submit an essay suggesting a potential namesake and why that person embodied the intrepid rover's spirit.

Over 3,500 entries were sent in. The eventual winner was Sojourner, suggested by 12-year-old Valerie Ambroise. Sojourner Truth was a freed slave, abolitionist and women's rights campaigner from the nineteenth century – a woman worthy of being immortalised forever amongst the stars. The word itself referred to a person who temporarily resides in a place. While there were no plans to bring the tiny rover home, it was hoped that it might forge the way for humans one day. The fit seemed perfect.

It was around this time that the Allan Hills meteorite made headlines around the world. Once again, Mars was at the top of everyone's agenda. While this was great for Pathfinder, it was too late for the MESUR mission that it was supposed to find a path for. Over the years, the ambitious sixteen-probe mission had been scaled back to just four, before having its funding pulled entirely. Rather than being the first station of a planetary network, Pathfinder would instead be the testbed for a new generation of Martian landers. All it had to do was land.

After a seven-month journey across the void, Pathfinder arrived at Mars on 4 July 1997 – exactly twenty-one years after Viking 1 was supposed to touch down. With the high-resolution Viking images to work with, there was no last-minute chaos trying to find a new landing spot. This time, the lander would make its Independence Day deadline.

However, there was still a hint of concern. Even these high-resolution images wouldn't see boulders a few metres across. If the lander bounced on a rock that big, it could puncture its airbag.

The landing selection team had used radar to find what they hoped was a nice smooth spot in Ares Vallis, just north of the planet's equator. As the team gathered in the control room back on Earth, the lander detached from the main orbiter. It struck the atmosphere at over 26,000km/h. The air resistance slowed the spacecraft down – first using a heat shield and then with parachutes – until the spacecraft was travelling a more manageable 225km/h. A set of retrorockets kicked in, slowing the spacecraft even more. Once it was almost at the surface, these thrusters cut out. Suddenly, the lander ballooned, literally, as the twenty-four airbags all deployed.

Eight seconds later, it struck the surface. Then it left again, rebounding 15m – the height of two double-decker buses – into the air. It bounced at least another fifteen times before finally coming to rest on its base.

A few minutes later, the spacecraft retracted its airbags, unfolded and radioed back to Earth. Pathfinder was on Mars and despite the bouncy touchdown, both landing station and rover were doing fine. The lander's first image was of Sojourner, just to prove the rover really was safe and sound. In the background lay a rock-strewn landscape, just waiting to be explored.

It all seemed too good to be true, and indeed it was. Within a day, the rover and lander stopped talking to each other.

The rover team frantically tried to find the source of the problem. There were concerns that the rover had entered silent mode, meaning it wasn't actively looking for signals anymore and could be permanently lost, but it should only do that when it was using the APXS to save power. It could be that the frequency of the cut-price radios had drifted too far, but they'd been working well until the signal dropped out.

Whatever the problem, the team soon found the solution – a good night's rest. At the beginning of Sol 2, the first downlink from the lander showed the rover was doing just fine.* It was time for Sojourner to leave its old friend Pathfinder behind and set out on its own.

The ramps, created by a company called Astro, resembled a giant metal tape measure, which unspooled. The lander had two ramps, pointing in opposite directions in case one was blocked. The rear ramp seemed more firmly planted on the ground, so the team decided that Sojourner should make its way to the surface backwards.

Knowing this was an important moment, imaging team member Justin Maki pointed the lander's camera at the ramp so he could create a short

---

*   The press, rather predictably, declared that the rover just needed turning off and on again, but in reality, the rover never reset itself. To this day, no one's really sure what caused Sojourner's brief temper tantrum.

movie of the rover rolling off.* For a while nothing came into view. Then, just as the team was beginning to get worried, a wheel appeared. A few moments later the rest of the lander followed. Finally, the first ever Mars rover had all six wheels on Martian soil. Sojourner was ready to put its geologist cap on and start investigating.

While it was continuously monitoring the Martian weather in the background, the static lander – now known as the Carl Sagan Memorial Station after the famous scientist who had passed away on 20 December 1996 – became a support station for the rover. It started by scouting the area, taking several panoramas to help select the best places for Sojourner to visit.

There were several large rocks nearby, surrounded by drifts of wind-marked sand. The rocks were quickly given nicknames from popular culture such as Yogi, Booboo, Roadrunner Flats, Zaphod and Indiana Jones. In the west were two mountains, named Twin Peaks, as well as several other features that helped the team back on Earth pinpoint the landing site on their Viking-based maps. Nearby, a collection of boulders known as the Rock Garden looked like a good place for a future trip.

When the first colour images came in, they revealed patches of darker soil where the retracting airbags had cleared away the top layer of bright red soil. There was even an arc of dark patches from where the lander had bounced.

The team began planning Sojourner's first drive. Her first target would be a rock near the rear ramp, named Barnacle Bill due to its many barnacle-like lumps. To plan out the route the team had to don some rather unusual (for the time) headwear – a pair of red-blue 3D specs, which made interpreting the images from the lander's stereoscopic camera much more intuitive.

Once Sojourner reached its target, the rover used its APXS to analyse the rock. Sojourner usually waited until after sunset to do this, using its precious reserves of battery power, as the night's chill reduced the noise, making the instrument more sensitive. To look at the physical properties of the rock, Sojourner had a special wheel that ground up against the surface. The cameras could then study how much material had rubbed away, revealing how hard or dense the rock was. The rover also spun its wheels

---

*   You can watch it here: mars.nasa.gov/resources/8776/sojourner-rover-rolls-down-the-ramp/

in the sands beneath it, watching the way the dirt moved. The experiment would go on to help design the tyres of future sister rovers.

By Sol 5, the rover had completed all its core experiments, not just at Barnacle Bill, but other rocks too and the rover was still going strong. The sols continued to tick by, and by Sol 30, Sojourner was still working as well as on Sol 1.

The same was not true of the ground crew. The team had been working flat out, assuming the mission would be over by now. They worked on Mars time, drifting from reality by 37 minutes each day, pulling 14-hour shifts and working through weekends. While that was manageable for a week-long mission, as the operations stretched out to a month the ground team began to revolt. The schedule was reorganised, and for the first time in a month, their day reverted to being 24 hours long.

On Sol 35, the rover began its way to the Rock Garden. The journey took several sols, as the rover could only travel a few inches at a time. On Sol 43, there was a moment of terror when the rover ended up precariously balanced on a rock called the Wedge. Sojourner stubbornly refused to move, its own safeguards telling the rover it was too dangerous. The operators overrode her programming, however, and got the rover down. The rover arrived at its destination on Sol 49 and began moving from rock to rock.

On Sol 56, sometime between midnight and 3 a.m., the on-board battery finally gave out. From now on, Sojourner would only operate while the Sun shone and would have to sleep through the night.

As time went on, Mars drifted further away from Earth, shortening the communication window with every passing day. It became harder to justify monopolising time on the Deep Space Network to keep the mission going. Not knowing how much longer they would be allowed to use the rover, once Sojourner was done at the Rock Garden, the team decided to see what the rover could really do. Up until now it had only driven inches at a time, but Sojourner was meant to be a technology demonstration; if rovers were going to do anything useful, they needed to be able to drive much farther than that. It was time to put the pedal down. On Sol 70, the rover drove away from the Rock Garden, traversing metres at a time, then looking back over its shoulder to look at the wheel tracks its speedy journey had left in the soil.

The mission's end came abruptly. The rechargeable battery on Carl Sagan Memorial Station was beginning to show signs of its age and was shutting down overnight to conserve power. On Sol 84, it didn't turn on again.*

Over its three months on Mars, the station took over 16,500 images, and 8.5 billion temperature, pressure and wind measurements. Meanwhile, Sojourner took 550 images, analysed sixteen different rocks with its APXS, undertook twenty-five mechanics tests of the soil and travelled a total of 104m. Yet, all this information came to just 287.5Mb – less than a single episode of your favourite show downloaded onto your phone.

In terms of geology, Pathfinder appears to have come down on the plains of some ancient catastrophic flood. The orbital imagery suggested the entire region was a flood plain. Pathfinder itself spotted rounded rocks, like pebbles worn smooth by a flowing river, and the stones of the Rock Garden were all aligned, suggesting they'd been dropped there by floodwaters. There were also signs of sedimentary rocks, built up over centuries by silt settling out of liquid water. Meanwhile, Sojourner's magnetic probes found dust infused with magnetic material. The most common way for this to happen is for iron-bearing minerals to dissolve in water, which then seeps into other rocks leaving these minerals behind. All in all, the evidence was piling up that Mars had once been a much wetter place.

If the rocks were borne in on the waters of a flood, then they didn't say very much about the location they had been found in. Instead, it was a grab bag of what was available across the planet. In bulk composition, these were largely the same as those found by Viking. The fact that three different landers in three completely different locations found that the rocks of Mars were almost all the same suggested that the Martian surface is far more uniform than our own patchwork planet.

While the rover was out poking rocks, the station had been keeping an eye on the weather. It seemed that even though the air was thin, wind had a significant effect on Mars, with several telltale signs of wind erosion etched upon the landscape. The station spotted several dust devils – tornadoes that were tens of metres wide and a few hundred metres high. One passed close to one of Pathfinder's solar panels, blocking out the light and

---

*   A few days beforehand, a contingency sequence had been uploaded to Sojourner. In the event it lost contact with the lander, it would drive around, circling the station until it received a signal. Who knows how long the plucky little rover circled around the station, waiting for a signal that would never come?

causing its power to temporarily drop. Evidently, the tornadoes contained a fair amount of dust.

Pathfinder also watched the thin atmosphere get thinner. It had landed during the planet's southern winter and on Sol 20 the polar ice cap reached its maximum size. As this cap is mostly made of carbon dioxide frozen out of the air, Pathfinder noticed a marked drop in the air pressure during these days.

The mission had been a great demonstration of how masterful NASA had become in garnering public interest. Within the first month, the Pathfinder website had 566 million hits. Bearing in mind this was in 1997, before the internet was considered a human right and most users were on dial-up modems, that is a very impressive number. Young audiences were being drawn in by the sight of a control room filled with young engineers dressed in jeans eating pizza with their feet up on the control panels rather than the starched white shirts and ties that had dominated human space-flight control rooms since the Apollo days.

Pathfinder's primary discovery, though, was highlighting the potential of Martian rovers. Sojourner had a few design flaws, but most of these were due to mounting an ambitious project under tight budgetary constraints. Despite this, the mission was lauded as a symbol of the incredible science that could still be done under the ethos of 'faster, better, cheaper'.

Unfortunately for Goldin, things were not going so well in other areas of the Mars exploration department. In the 1998 launch window, two other low-cost spacecraft were due to fly to Mars, forming the Mars Surveyor 98 mission. The first was the Mars Climate Orbiter, costing just $125 million, which would explore the planet from above, while also acting as a relay station for the second spacecraft, the Mars Polar Lander, that was due to arrive a few months afterwards.

The Mars Climate Orbiter was scheduled to enter orbit on 23 September 1999, but its final course correction went wrong. The reason for the error: someone had failed to convert a single measurement from the imperial units used by contractor Lockheed Martin to the metric units used at NASA. In an attempt to meet the deadlines set by 'faster, better, cheaper', the code hadn't been checked quite thoroughly enough. It was an all-too-human mistake that remains the butt of jokes even two decades later.

It was hoped that the Mars Polar Lander would fare better when it arrived on 3 December. It was due to come down in the planet's southern polar region, to investigate how the poles changed with the ebb and flow of the seasons.

To avoid another embarrassing failure, NASA triple-checked the code for mistakes. The descent began just as planned. It was unable to keep contact with Earth during landing and the probe's radios cut out. The plan was for the signal to reconnect once it arrived on Mars. It never came. The team listened for several days but heard nothing. The lander is thought to lie somewhere in the Planum Australe region, although in how many pieces is anyone's guess. As there was no information from the landing, no one knows when the mission went wrong.

The double failure was an embarrassment, turning NASA into a punch-line for the likes of Jay Leno to poke fun at. 'It proves,' the comedian said on his late-night show, 'that you don't need to be a rocket scientist to be a rocket scientist.'

The double failure shook Congress's faith in NASA and Mars was not the only wing of the agency experiencing problems. The new International Space Station had over run by billions while a proposed reusable spacecraft, the X-33, was about to be abandoned after $1 billion of development. It was time to get to the bottom of the problem, and NASA was subjected to an extensive review.

'Over the past year, I have been continually amazed by reports coming out of NASA about the mission failures and program delays,' said Senator John McCain while heading a meeting of the Committee on Commerce, Science and Transportation on the subject. 'The extent of the mismanage-ment noted in these reports is somewhat startling.'

During the review of the Mars programme known as the Young Report, the finger was pointed at everything from shoddy workmanship from contractor Lockheed Martin to President Clinton's lack of funding. Ultimately, the fault wasn't financial or mechanical, but managerial. The project had pushed too hard too fast.

The principle of 'faster, better, cheaper' unarguably had its benefits. However, the zealousness with which it was instituted proved to be its downfall and Goldin's too. He stepped down as administrator in 2001. It was time for a new NASA.

17

RED ROVERS

The year 2000 was a strange one for NASA's Mars programme. The excitement for Mars whipped up by Pathfinder and the discovery of the Allan Hills meteorite (even though it now looked like its fossil bacteria were actually just an artefact of the imaging process) was still there, but there were serious doubts over the direction Mars exploration should take. With the iron grip of 'faster, better, cheaper' slipping, there were now fears that NASA would swing back the other way to over-bloated, expensive missions.

Congress called a hearing on 20 June 2000, a few months after the release of the Young Report, summoning experts to give their thoughts on how NASA should proceed to Mars. Among them was Louis Friedman, head of NASA's Mars programme during the Viking years and now executive director of the Planetary Society, a public advocacy group for solar system exploration he'd co-founded alongside Carl Sagan and Bruce Murray. He said:

> The public wants to continue on the road to Mars and they strongly support learning from recent mistakes in order to resume the exploration of our sister planet. The recommendations that strongly emphasise technological development could unnecessarily lead to NASA's scaling back too much on its basic objective of scientific exploration. The trick is to find the right balance between more careful spacecraft development and operations on the one hand, and on sustained discovery and progress on the other.

Congress and NASA agreed. The key was running a mix of missions with a range of budgets. These missions would be based around one central tenet: follow the water. Water is one of the key ingredients of life, at least as we know it. Wherever on Earth you find water, you find life. The public's love of Mars was still tied to the hope of finding life, but Viking had shown explicitly that hunting for biology was a dangerous gambit. By following the water, NASA would be able to slake the public's thirst to hunt for Martian life without promising to find it. Whatever mission came next, water would be its goal.

There was plenty of evidence from Viking, Pathfinder and the many orbiters that water once flowed across Mars. There were even traces of vapour in the atmosphere and signs of ice frozen into the soil. Now it was time to understand the history of that water – what it was like and where it went. The agency came up with a decade-long plan to explore Mars, one that would launch a new mission to the planet every launch window.

The first of these would come in 2003, and NASA was keener than ever to take advantage. The orbits of neither Earth nor Mars are perfectly circular, meaning sometimes the distance during opposition (when the planets are closest) varies, and can be almost double what it is at other times. In 2003, Mars would pass just 55.8 million km from Earth, its closest approach until 2018. With a shorter distance to travel, NASA would be able to get a lot more bang for their buck in terms of fuel, meaning they could fling a much heavier spacecraft at Mars. But it would be a risk. If they sent a bigger mission they would have to meet their deadline, as the rocket wasn't large enough to send it the extra distance come the next launch window.

The obvious choice was a rover. As it was mobile it would be able to literally 'follow the water', moving across the surface to the places it would be most likely to find evidence of Mars's wet past. Since the late 1990s, the agency had been planning a bigger, beefier version of Sojourner. They'd already worked out the trickiest part of the mission, the landing procedure. All they needed to do was upscale their airbags to deal with a heavier rover.

The plan quickly hit a snag. Initially, the rover was designed to be huge so it could carry every instrument and experiment possible, but this would require making the tetrahedral lander box larger. This turned out to be a much harder prospect than anticipated and the team couldn't make the changes before the 2003 deadline. The box would stay as it was. The rover would have to be the one to change.

Normally in spacecraft it's weight that governs what goes to space and what stays on the workbench. Now the team faced another rule: will it fit? For stability, rovers are designed to be long and squat – exactly the wrong shape to fit in a pyramid. If the rover was going to fit, then engineers were going to have to fold it up and cram it in.

Talk to any space engineer and use the word 'moving parts' with reference to their mission and you'll see them break out in a sweat. Moving parts go wrong. Lens caps seize.* Folded antennae get stuck.** Cables fail to detach.*** But if NASA wanted to send a big rover in 2003, this was the only way.

There would be a few compromises in terms of science. Rather than the comprehensive machine they'd wanted, the rover would have a limited number of instruments, refocusing it as a robo-geologist.

All this design work was being done speculatively. In July 2000, a panel of judges came together to decide what direction NASA's planetary exploration would take. The team from the rover were pitted against the Mars Science Orbiter (MSO), which would study the planet from above.

There was no debate that the rover was the 'sexier' of the two projects. Sojourner had shown how charismatic a rover trundling around the surface could be and NASA runs as much on public opinion as it does on science or taxpayers' dollars. But it was by far the riskier of the two. NASA was still recovering from the twin failures of the Mars Climate Orbiter and the Mars Polar Lander and refused to chance another high-profile failure.

Arguing the case for the rover was Principal Investigator Steve Squyres. He'd been trying to get an instrument on a Mars mission for almost fifteen years, but bad luck, and scaled back and cancelled missions meant he'd never quite managed it. Now he was in charge of not just an instrument but an entire rover. Would this be another disappointment? Halfway through the debates, Squyres received a call from Goldin's office (who was

---

*   Venera 10 and 11.
**   Galileo, a 1989 probe to Jupiter, had an antenna that was supposed to open like an umbrella. It didn't.
***   This happened to BLAST, a balloon-borne telescope. In 2006, its parachute failed to detach after it landed, and the telescope was dragged 200km across the Arctic tundra, scattering wreckage. They did locate the hard drives, but it took several months as no one had thought to change their colour from white.

still administrator in 2000). Squyres braced for bad news. But the question he got was entirely unexpected: 'Can you build two?'

It had been a long time since NASA had been in the habit of doubling up to avoid failure, but the logic was still sound. The chances of both rovers failing was much less than for one. But after years of 'faster, better, cheaper', the idea of NASA offering money to increase the scope of a mission seemed ridiculous. After a few hours of back-of-the-envelope calculations, Squyres got back to them – one rover would cost $440 million, two $665 million.

It took the panel several weeks to come to a decision, but ultimately the call came down. The rovers, now called the Mars Exploration Rovers (MER), would be the ones heading to Mars in 2003.* Squyres had three years to achieve his dream of setting down hardware on Mars and build not one, but two rovers.

The rovers had four main goals: find out if life could have existed on Mars; characterise the planet's past and present climate; and investigate the geology and prepare the way for human exploration.

The first three queries were centred around finding whether there was, or still is, liquid water on Mars. They would be tackled by newly updated versions of the same instruments that had gone on previous rovers and landers. Cameras would survey the landscape; a microscope would take a closer look at the soil while an APXS measured its composition; and a rock abrasion tool would help to buff away the weathered outer layers of boulders.

The spectrometers, microscope and rock abrasion tool would all be carried on the end of a robot arm so they could get a close look at their target. Improvements in lightweight but strong materials meant the arm was much slenderer than Viking's had been. It also bore a hinge halfway down its length, greatly increasing its manoeuvrability. It was an elegant design, and one that has been largely unchanged for all the Martian landers that have followed. Together, these would hunt out those minerals and rock formations that occur in the presence of liquid water on Earth, meaning it was likely they did on Mars, too.

---

*   The MSO did end up flying to Mars in 2005 as the Mars Reconnaissance Orbiter.

The last point – paving the way for humans – was a more esoteric goal. Landing a human on Mars was still a long way off, but the mission could hope to glean what resources might be available on the planet that a future explorer could exploit.

As development progressed, it wasn't long before the problems started mounting, beginning with the solar panels. The minimum lifespan NASA expected from the rovers was just 90 sols. As ever, the engineering and science teams wanted them to last as long as possible. Provided the rover was built robustly enough, the main culprit for ending the mission was likely to be the power supply. During its mission, the Sojourner rover had suffered from a steady drop in power as dust in the atmosphere settled on its solar panels, blocking the Sun. The same would happen for the MERs until they were covered in so much dust that they wouldn't be able to charge enough to keep warm through the night.

To keep that eventuality as far in the future as possible, the solar panels needed to be as big as was feasible. The size of a solar panel is measured in strings – a number of individual elements linked together. To reach the 90-sol working lifespan, the rover would need thirty strings. The initial plan was to have these thirty strings on several triangular panels, which would unfold to form a hexagonal plate across the top of the rover. However, Squyres wanted to get as much out of the MERs as he could. Was there a way to squeeze in a few extra strings?

He farmed out the problem to engineer Randy Lindeman, who spent months toying with designs, eventually striking on a winner – adding two small triangular winglets. These would allow the rover to have thirty-six strings, which had the potential to drastically increase the length of the mission.

The problem was, the winglets added to the weight of an already bloated rover. The rover had managed to put on 15kg that it now it needed to shed. The team scraped away what they could, but ultimately something big had to go. Either they could take off the winglets or remove the ramps the rover used to get off the landing platform. While the rover could make it to the surface without them, there was a risk it would flip over. Squyres fought for the winglets, arguing they'd regret their loss when they got to Mars. However, if they lost the ramps, they risked ditching the rover in the dirt on Sol 1. It wasn't worth risking the 90 sols they could have for the extra days they might have. The winglets had to go.*

Then came the kicker. The vendors supplying the solar panels said they couldn't fit the thirty strings they'd promised on the base design, only twenty-seven. Even the winglet version could only fit thirty-four strings. So, back on the winglets went. Instead, they had to solve the weight problem the old-fashioned way – by throwing money at it. With extra funding, the fabricators could replace the cheap but heavy components with light but expensive alternatives. The question then was where the money would come from.

One solution would be to lose the second rover. That would free up nearly $200 million, but that was an option no one wanted. NASA managed to cobble together enough funding to cover the overrun but that would be it. If anything else went wrong, the second rover would have to go.

While the rovers were being built, the entry, descent and landing (EDL) team were looking for a good spot on Mars. Those big solar panels would be useless without sunlight to charge them, and the most reliable sunshine was found around the equator. Next to consider was elevation. If the lander came down on a high mountain range, it risked hitting the ground before the parachutes had the opportunity to slow it down. Finally, the site had to be scientifically interesting and likely to contain evidence of water on Mars, either past or present.

Three spots met those criteria: Melas Chasma, Meridiani Planum and Gusev Crater.

The most alluring was Melas Chasma – a giant canyon that stretched for thousands of kilometres. There was a good chance of seeing layers of exposed rock. As these layers had been laid down over successive millennia, to geologists they were a storybook of Mars's history. Unfortunately,

---

*   The need to match mission requirements versus the desire to exceed them is another recurring friction of spacecraft design. Administrators set down the minimum limits to keep the project on task and on budget, but everyone involved in actually building and using the spacecraft wants them to be as good as possible. The design teams at NASA frequently push the laws of both physics and economics to do so. Given how often NASA spacecraft outlive their planned lifespan by years if not decades, I think we can all see who usually wins that fight.

it was also home to strong winds. Combined with an inaccurate landing system, there was a risk the lander would get smashed into a canyon wall. So Meridiani and Gusev it was.

Now, the EDL team needed to make sure the airbags could survive the extra weight of the rovers, so they grabbed a set of airbags (once again made by the seamstresses of ILC Dover) and took them to the world's largest vacuum chamber to test them. The airbags failed. Completely.

With the extra weight, rocks ripped straight through the airbags. Frantically, the EDL team designed thicker bags, adding a second bladder in case the first failed. The changes made the bags heavier on a system that was already pushing its weight limit but eventually they reached a point where the bags would cope.

The EDL team was confident the bags would manage on the smooth and flat Meridiani Planum, thought to be a former flood plain. The problem was Gusev Crater. For the last billion years, the crater had been bombarded by space rocks and eroded by Mars's weather. It was covered in spiky rocks ready to rupture the first unsuspecting airbag to bounce on them. To make matters worse, the crater's winds could drag the lander over the jagged edges.

If the wind was going to blow them off course, then they were going to have to just blow right back. They added a new system (Descent Image Motion Estimation System, DIMES), which took pictures of the ground below the lander, judged how far the wind had blown it and then used thrusters to compensate. After several missions whose first pictures had revealed a nearby boulder that might have killed the mission, this small degree of control was a welcome addition.

The rovers, just as expected, had captured public attention, helped in no small part by the fact they looked adorable. The twin lenses of the spectroscopic camera looked like eyes on a cartoonish face, perching on a long, neck-like mast – a look that years later would serve as inspiration for the lovable protagonist of Pixar's *Wall-E*. Meanwhile, the gold foil-wrapped main body of the rover was the same 'a box on six wheels' design as Sojourner, albeit three times larger. On top of this were the solar panels. Once they were folded out, these formed a swept-back V-shape, reminiscent of a fighter jet. With their robotic arms reaching out, it was all too easy to think of them not as robots, but sentient explorers each with their own personality.

Rovers that were this cute needed a name. Just like Sojourner, the rovers (again female) were named by an essay contest. The winner this time was Sofi Collis, a 9-year-old who spent her early years in Siberia before moving to the United States. 'I used to live in an orphanage,' her essay read:

> It was dark and cold and lonely. At night, I looked up at the sparkly sky and felt better. I dreamed I could fly there. In America, I can make all my dreams come true. Thank you for the 'Spirit' and the 'Opportunity'.

Spirit and Opportunity seemed the perfect encapsulation of a mission that would send a pair of explorers to a strange new world. Rover missions have, and continue to be, much more than tele-robotic-geologists. They go where humans cannot, a symbol of our quest to touch the worlds around us, to learn and to explore. They are symbols of our spirit, and the things we can achieve when we set our minds on them. They are symbols of the opportunities that lie within our grasp if only we dare to reach for them. They are a symbol of hope for a future that is brighter than our past.

This was no more eloquently highlighted than in two small, seemingly insignificant, parts hidden away on both landers. The rock abrasion tool, which would grind away the surface layer of rocks, was being crafted at Honeybee Robotics, a company based in lower Manhattan. On the morning of 11 September 2001, as the employees were making their way in to work a boom ripped through the entire city. A plane had crashed into the World Trade Center, just a few blocks away.

As the weeks past, the team attempted to return to work as normal. But, how could they? They had been on the doorstep of US history's greatest tragedy. There was little they could do to help clear the rubble or ease the suffering of the victims' families, but what they could do was send a memorial to Mars.

After contacting New York Mayor Rudy Giuliani, Honeybee were gifted a few twisted fragments of aluminium from the wreckage of the Twin Towers. They were crafted into cable shields, protecting the electrical wires from damage. Each bore a US flag. When the rovers landed on Mars, they would take a part of the United States' soul with them.

As the mission progressed, the two rovers began to take shape. In theory, both Spirit and Opportunity* should be identical, but it quickly became apparent Spirit was a problem child.

The first sign of something untoward came when Spirit was undergoing a vacuum test. The rover refused to co-operate. The flash memory that served as its hard drive was playing up; several commands weren't operating correctly. In the end, the Flight Software team created the SHUTDOWN_DAMMIT command to get the rover to shut down, dammit, hoping the ancient ritual of turning it off and on again would work.

Elsewhere on the rover, the stereoscopic camera, PanCam, was producing images so blurred by speckling they were completely useless. The culprit was a single wire made out of the wrong material. They replaced the cable but ran out of time for testing – they'd only know if it worked when they got to Mars.

That's not to say that Opportunity was a perfect sibling. One of the sensors on the APXS was cracked and had to be replaced. Everything was fine until instrument testing, when the smell of burning filled the lab. Something had shorted, cooking not just the instrument itself but the test equipment, too. The culprit was found to be a single sliver of metal caught in an electronic pin. There was no way to fix it. They would have to use the flight spare. While created to the same standard as the 'official' instrument, it wasn't nearly as well calibrated but there was no other option.

The biggest problem for the rovers came in the eleventh hour. In February 2003, the rovers were transferred from JPL to Cape Canaveral for final assembly before being installed on the Boeing Delta II rockets that would carry them to Mars. A few months later, during testing they discovered a flaw in the Telecom Services Board – the electronics that control the rover's communications systems, located deep inside the rover. It was a relatively easy repair to make on Opportunity, but Spirit had already been packed up and stowed, ready for its flight.

It had been trussed up with cables that were meant to prevent it shifting during flight, all of which would now have to be cut. On Mars, this would be done using pyros – small gunpowder charges that fire a blade to cut through the wire. To release Spirit, they fired off all the pyros, forgetting

---

*   During their manufacture, the rover that became Opportunity was referred to
       MER-1, while Spirit was MER-2. However, as Spirit launched first, it was called
       MER-A, while Opportunity was MER-B. That's all very confusing, so I'm just
       going to refer to them by their popular names.

these had been crafted to work in the thin atmosphere of Mars. As they blew the pyros, they also blew a fuse.

The fuse in question had been installed to protect the more delicate instruments during testing, so wasn't critical to the rover's operation. Even with the fuse blown, the rover worked just fine and was given a green light to fly without replacing it. The question now was what had caused the fuse to blow in the first place, and would it happen again on Mars with more serious consequences?

The culprit was the pyros. After detonation, they released a cloud of conducting gas that allowed a current to flow where it shouldn't. It was only for 32 milliseconds, but it was long enough to fry the fuse. It was also enough to fry the resistors responsible for firing the pyros, potentially preventing them from cutting the cable. If that happened on Mars, the rover would be trapped. The only way the directors would let either rover fly was to find all the pyros and prove they had deployed correctly.

No one had expected to need the pyros ever again. On 21 May, with only nine days left until Spirit's launch window opened, the team began searching high and low for every single pyro. If they didn't find them, the mission could be pushed back until 2005, if it wasn't cancelled outright.*

As they hunted, the launch team continued to mount Spirit on its rocket, even though the mission was grounded until the pyros were found. One by one they appeared, all of them had detonated just as they should have. Finally there was just one missing. Out on the launch pad, Spirit's Delta II was being fuelled up, ready to carry the rover to Mars. If they could just find that one last pyro …

Then, in a bag, stuffed inside another bag, buried at the bottom of a locker, they found it. It too had detonated. The mission was cleared to launch.**

While Spirit managed to get under way without issue on 10 June – watched by VIPs including Sofi Collis and one of the seamstresses from Dover who had made the airbags – Opportunity had a rougher start. The first launch attempt had to be scrubbed due to bad weather and

---

*   Ed Weiler, Associate Administrator of the Office of Science, threatened Steven Squyres with turning the two rovers into museum exhibits, which has become the fate of several completely serviceable space missions that never flew after falling victim to budget or political caprice.
** As with most missions, the rovers had launch patches designed for them. Spirit's bore the Loony Tunes character Marvin the Martian, while Opportunity's featured Daffy Duck in his Duck Dodgers guise.

was pushed back to the next day. Only, the next morning when the launch team at Canaveral checked the rocket, they discovered the cork was peeling from the outside.

You read that right. Cork. A thin layer of the wood covered the outside of the Delta rocket, helping to spread out the heat caused by friction during the launch. When Boeing, the company that had made the Delta II's, assembled Opportunity's, they had stuck the cork down incorrectly, allowing water to seep in and causing the layer to peel away. A team from Boeing re-stuck new panels but the cork peeled away again. After several frantic phone calls with the glue company, Boeing finally managed to get the cork stuck back on and the rocket launched on 8 July.*

NASA had wanted to find a balance between pushing forward the knowledge and technology of space science while not being swept away by rapidly expanding costs. Although their construction had gone over budget, NASA had two rovers to show for their efforts, both of which would be able to go farther and look deeper than any Mars mission had gone before. It was time to get them to Mars.

---

*   As the launch was running several days late, only a few of the rover's creators had stayed behind. Among them were Mary Mulvanerton and Jon Beans Proton, both of whom were bagpipe players. As Opportunity rose into the sky, it was accompanied by their comforting drone to the strains of 'Amazing Grace'.

18

SPIRIT AND OPPORTUNITY

During the flight, NASA set about preparing for the long road ahead. The furthest the agency had ever driven a planetary rover before was the 100m that Sojourner had traversed. The MERs were expected to traverse ten times that distance and there would be two of them operating at once. Throughout the journey to Mars, the Ground Control team ran drills with test rovers, crafted to be identical to Spirit and Opportunity, making sure that when the real rovers arrived, they could dive straight into the task of investigating Mars's wet past.

In the meantime, there was a brief respite ten days before Spirit was due to arrive when another agency's spacecraft approached the Red Planet. The European Space Agency (ESA) were also taking advantage of the close pass of the two planets to send their first ever Mars mission, Mars Express.

The agency was set up in 1975, pooling the space exploration efforts of more than a dozen nations, knowing their combined strength would be far greater than the sum of its parts. Yet, even with their combined finances, ESA's resources paled in comparison to NASA's. Today, NASA spends more on human and robotic exploration in three weeks than ESA does in a whole year.

However, the Europeans didn't let this financial discrepancy dull their ambitions, undertaking their own, more modest, planetary missions. One way was getting individual instruments to pay their own way, which was how Mars Express came to carry a lander called Beagle 2.*

The idea for Beagle 2 had been germinating in the mind of Colin Pillinger, a researcher from the Open University and member of ESA's Exobiology Study Group, for years. Pillinger built his lander on just £50 million – less money than Spirit and Opportunity had overrun by. He'd raised the money almost by sheer charisma alone, persuading people to donate their work for cost while also wrangling various grants and donations before convincing ESA to carry the lander on Mars Express.

It was a modest spacecraft, managing to cram a lot into a tiny weight budget. It would hunt down the carbon-bearing organic molecules that all Earth life is based on in Mars's soil and atmosphere. Particularly, it would sniff out methane. Although the gas can be released by geological processes, it could be a sign of Martian microorganisms in the soil, and so exobiologists were keen to see if there was any methane in Mars's atmosphere.

Beagle 2 quickly made a name for itself in the UK. Not only was it a British-built spacecraft hitching its way to Mars, Pillinger's broad West Country accent made him a likeable and memorable figure, regularly appearing on TV to stoke enthusiasm for the project. Unfortunately, this enthusiasm was a set up for failure. As the lander descended towards Mars on Christmas Day in 2003, it had to run silently. Pillinger spent the holiday waiting for it to make contact with Earth from the surface but heard nothing. The Beagle 2 team listened for several months, finally giving up the spacecraft as lost in February 2004.

The lander's fate remained shrouded in radio silence until 2016, when the ultra-high-resolution Mars Reconnaissance Orbiter snapped an image of it on the surface. Only four of Beagle 2's five solar panels had unfolded properly, meaning the transmitter was still covered by the last, leaving it unable to phone home.

A model of Luna 9, the first spacecraft to softly touch down on the Moon. During landing, the top half of the spacecraft released the egg-shaped lander, cushioned by an airbag. (Ezzy Pearson/Science Museum London)

An engineering model of Lunokhod 1. The rover is pictured in its 'night' state, with the solar panel covered to conserve heat. The yellow drill-like appendage is an antenna, while the 'laser gun' is an X-ray telescope to study background radiation. (Ezzy Pearson/Science Museum London)

Left: A model of the Venera 7 lander, the first spacecraft to transmit from the surface of Venus. The lander was suspended below a parachute that opened as it descended through the atmosphere. (Ezzy Pearson/Science Museum London)

Below: Colour image of Venus's surface sent back by Venera 13. There is a lot of debate about the correct way to colour-balance these images to replicate what the surface of Venus accurately looks like because the atmosphere filters out blue light. While this image appears very yellow, other versions appear more neutral. (Don P Mitchell/Russian Space Agency)

Astrophysicist Carl Sagan poses with a scale model of Viking for his television series *Cosmos*. Sagan was one of the leading voices in explaining planetary exploration to the public. (Cosmos-A Personal Voyage/Druyan-Sagan Associates, Inc.)

Three generations of Mars rover (front left: Sojourner; back left: Mars Exploration Rover; right: Curiosity). Over the years the rovers have grown in scale. These models are the flight spares, with Sojourner's spare named Marie Curie. (NASA/ JPL-Caltech: mars.nasa.gov/resources/3792/three-generations-of-rovers-in-mars-yard)

Pathfinder's view of its landing site. Sojourner is in the centre of the image, investigating a rock known as Yogi. (NASA/JPL: www.jpl.nasa.gov/spaceimages/details.php?id=PIA01466)

The hematite 'blueberries' discovered by Opportunity. The mineral forms only in the presence of liquid water, suggesting Mars was once much wetter than it is today. (NASA/JPL-Caltech/Cornell University: www.nasa.gov/mars_art/blueberries#.XkRCKxP7QW9)

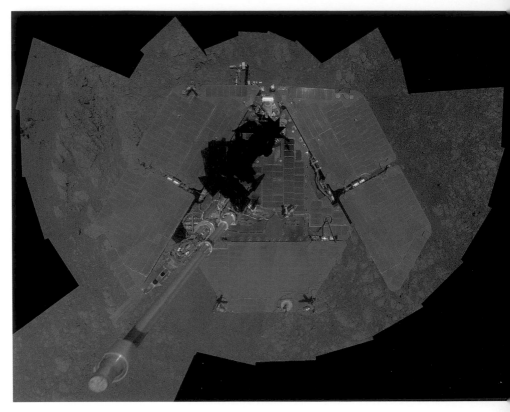

Two of Opportunity's selfies. The left image was taken in January 2014, and the right in March 2014, after wind events wiped the solar panels clean. The black region in the centre of the image is the position where the camera is located. (NASA/JPL-Caltech: www.jpl.nasa.gov/spaceimages/details.php?id=PIA18079)

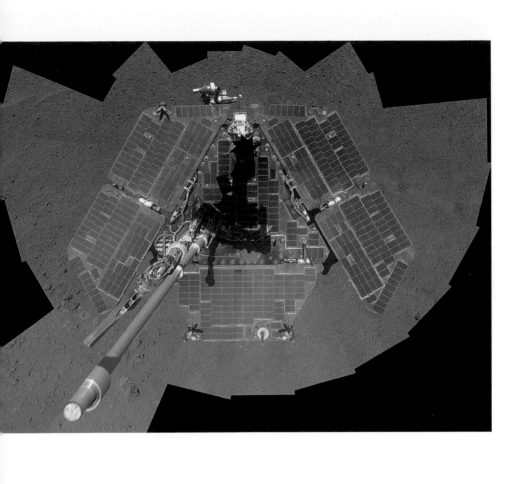

Engineers testing the Phoenix Lander ahead of its deployment on Mars. The same design would be used a decade later for the InSight Lander. (NASA/JPL/UA/Lockheed Martin: www.jpl.nasa.gov/spaceimages/details.php?id=PIA01885)

Self-portrait of Curiosity taken on 11 October 2019 while at Glen Etive, a location in the clay unit of Mount Sharp. (NASA/JPL-Caltech/MSSS: www.jpl.nasa.gov/spaceimages/details.php?id=PIA23378)

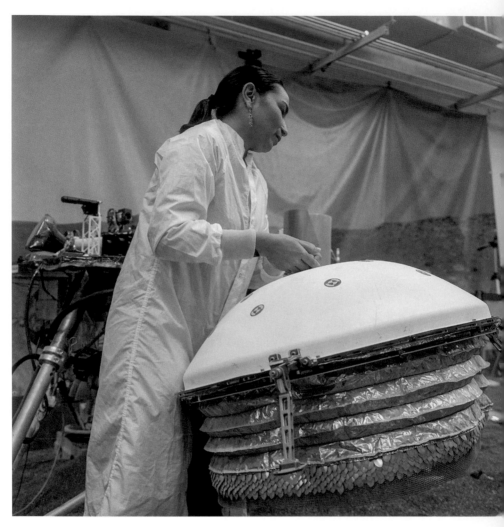

JPL Systems Engineer Marleen Sundgaard holding the wind and thermal shield for the SEIS instrument on InSight. (NASA/JPL-Caltech: www.seis-insight.eu/en/public-2/seis-instrument/wts)

The view taken by Hayabusa 2's mini lander, MINERVA-II, just after it arrived on the surface of asteroid 162173 Ryugu. The image was released unprocessed, which is why it looks distorted and has lens flare. (JAXA)

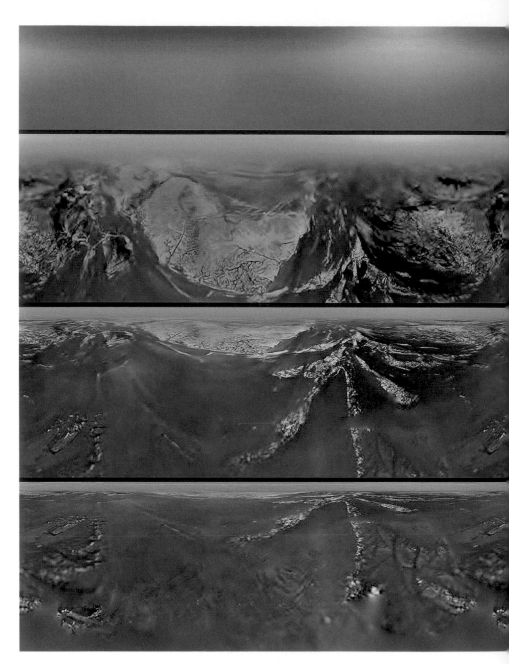

The view of Saturn's moon Titan as the Huygens probe descends through the atmosphere. The top image was taken while the probe was still above the thick cloud that covers the surface, while the bottom one was taken shortly before landing. (ESA/NASA/JPL/University of Arizona: www.nasa.gov/content/ten-years-ago-huygens-probe-lands-on-surface-of-titan)

The Chang'e 3 lunar lander looks out at the Yutu lander as it makes its way across the lunar surface. (Chinese Academy of Sciences/CNSA/The Science and Application Center for Moon and Deepspace Exploration/Emily Lakdawalla: www.planetary.org/multimedia/space-images/earth/yutu-on-the-road.html. More raw images of Yutu and Chang'e can be found here: planetary.s3.amazonaws.com/data/change3/pcam.html)

Mission Control during the landing of Indian lunar lander Vikram as part of the Chandrayaan 2 mission. (ISRO)

In 2004, it was a timely reminder to NASA of how hard the Curse of Mars had hit them previously, a thought they were now forced to endure for the ten-day wait between Beagle 2's failure and Spirit's arrival at Mars. The rover came in for its landing on 4 January 2004. Far from the horror of Beagle 2, Spirit landed perfectly. The airbags didn't pop,* the pyros fired, and the rover began to slowly unfold itself. Eager to get a look at its new home, Spirit deployed its camera and took its first panorama. Just as the EDL team had planned, they had come down on a former lakebed, the perfect place to look for evidence of past water and investigate Mars's climate.

As with every Martian landing site up to now, there was a scattering of boulders but generally the going was flat and easy. After a few sols dealing with a snagged airbag, the rover finally got its wheels on Martian soil on Sol 12, or 15 January to us Earthlings, ready to begin its work investigating Mars's wet past.

As the rover began making its way across the surface, its images were coming in crisp and clear with none of the speckling that had plagued early tests. It seemed that all the problems of Spirit's youth were gone. And they stayed gone for a whole 5 sols.

On Sol 18, Spirit stopped talking. The next day, it sent a message that it was in fault mode as the computer constantly rebooted itself. No one could find out why because the rover was stubbornly refusing to stay operational long enough to download the logs that would tell them what was wrong with the rover. Even the SHUTDOWN_DAMMIT command wasn't working. Spirit was out of control.

The only option was to hope the software was at fault, because if it was a hardware issue they couldn't exactly send a mechanic to fix it. The programmers began going through every line of code, looking for the cause of the issue. Finally, they found the problem – the flash memory that had given them so much trouble during testing wasn't clearing properly and was now full. Their only hope was to spam the rover with commands to reboot without the flash drive and hope one got through.

---

*   When the signal came in to say that Spirit had survived its first and biggest bounce, the cheer in the auditorium was so loud that the EDL team had to shout to each other so they could actually finish the landing.

It worked. On Sol 21, the rover began communicating with Earth again, allowing them to deal with the symptoms even if Spirit's underlying illness remained undiagnosed.* All they could do was hope the rover would behave.

Meanwhile, Opportunity was coming in for its own landing, touching down in the Meridiani Planum on 24 January. When its first images came in, the watching technicians fell silent. It seemed like the 'land of Opportunity' would be a Mars unlike any seen before – a smooth desert, where the only real features were the dark marks left behind by the airbags.

As more of the scene was revealed, the camera exposed geological pay dirt (literally); an outcrop of layered rock directly in front of the rover. They had come down in the middle of an impact crater – later named Eagle Crater. It seemed that the impact had punched through to the underlying bedrock. They'd finally be able to look at some stone in the place it had been created billions of years ago.

There was something equally interesting in the form of small spherical rocks, which became known as 'blueberries'. These appeared to be hematite concretions – mineral balls created by water dripping into round rock cavities. Over time, the soft surrounding rock gets worn away, while the hardy hematite balls remain behind. They were everywhere within Eagle Crater, earning it the moniker 'The Blueberry Bowl'. NASA had come to 'follow' the water, only to find evidence of it before barely making it off the landing platform.

Opportunity was once again the charmed twin, but its luck didn't last long. On Sol 2, the team discovered that a heater on the spacecraft's robotic arm was stuck on, draining their meagre power supplies. There was enough power for the rover to keep running normally, but it did mean that Opportunity was likely to suffer an early death.

With only a limited lifespan, time management became king. It would be all too easy to spend the entire 90-sol mission just investigating Eagle Crater and its outcrop of bedrock, now called Stone Mountain.** But the crater was shallow, meaning only the top few layers of Mars's most recent geologic history were on show. The rovers were built to go the distance.

---

*   In the surface mission support area back on Earth, a white board proclaimed the
        closest thing they had to a diagnosis for the rover – 'The Spirit was willing, but the
        flash was weak.'
**   A rather misleading name, seeing as how the outcrop was just 10cm high.

For all the great science to be done in Eagle Crater, there could be an even greater site just over the horizon, such as the 130m wide Endurance crater just 1km away.

Over in Gusev, the Spirit team were having similar issues deciding how best to spend their time. Judging by how much dust was settling on the solar panels, the team estimated Spirit probably had around 220 sols in her, far longer than the designed 90 sols. Opportunity, with its power-sucking heater, would probably last less. 'It would be easy for us to get bogged down,' said Squyres. 'We need some kind of goal to push us on towards Endurance. So I'm going to tell the team that I want us to be there by Sol 90.'* The science team complained at the deadline, but the decree forced them to prioritise only the most interesting investigations.

Opportunity reached Endurance on time and began to hunt around the crater's lip, tracking down a place to creep inside. While the rover could manage a downward slope of 45°, it could only climb back up 20°. While it was likely that Opportunity's solar panels would give out inside the crater, NASA wanted the option of continuing the mission afterwards. They found a shallow enough slope, and on Sol 132 Opportunity cautiously made its way down.

Spirit soon received its own deadline – it was to strike out for the distant range of low hills, the Columbia Hills. But, where Endurance had been a journey of less than a kilometre, these were 2.7km away. Spirit hiked across the landscape in stops and starts – travelling around 100m every morning before the rover stopped to recharge, but two months later Spirit reached the hills' base.

The sols passed, and dust began to build up on the solar panels. When Spirit landed in January 2004, it was generating around 900 watt-hours per sol. By December that year, it was down to just 400 watt-hours. Then something unexpected happened. Once again, Opportunity was the lucky twin. At random intervals throughout the year, the rover's energy output jumped up by as much as 5 per cent overnight. During the night, wind currents were whirling up around the rover, blowing the dust off the panels. By the end of 2004, Opportunity was generating

---

*   S. Squyres, *Roving Mars: Spirit, Opportunity, and the Exploration of the Red Planet* (Hyperion, 2005) p.321.

more power than its twin. By March 2005, the wind was cleaning Spirit's solar panels as well.

Dust build-up had always been the main life-limiting factor for the rovers. Now the rovers were newly refreshed and rejuvenated. They could last for months, if not years. With their horizons suddenly expanded, it was time for the operators to start thinking bigger.

They'd already driven 2.7km to Columbia Hills without a problem but there were other hills to explore; other craters, too. However, the rovers hadn't been designed to survive the approaching Martian winter, so the teams tracked down places where the pair could hunker down, with enough sunlight to keep them charged and warm through the long nights to come.

The missions continued for over a decade. Spirit strode across difficult and steep terrain, spending much of its time in a region known as Home Plate – a wide volcanic deposit. Opportunity, meanwhile, began crater chasing, road-tripping from Endurance to the larger Victoria Crater before finally reaching the 22km-wide Endeavour. The rovers received five mission extensions, outperforming the wildest hopes of the engineers who had created them.

But no man lives forever and neither does a robot. Predictably, troubled Spirit was the first to go. In May 2009, after roughly 1,900 sols on Mars, the Spirit rover wheels began having problems gaining traction. To make matters worse, one of Spirit's wheels had failed in 2006, and the rover had spent the last three years driving backwards, dragging its limp foot as it went.

This wasn't the first time a rover had got stuck; Opportunity had done so in 2005 but wrestled itself free. The team were confident they could do the same for Spirit. The problem was that the Spirit's belly was pressed against a rock, preventing its wheels getting purchase in the sand. Spirit tried everything she could to get out of her predicament. After months of trying, NASA made the call in February 2010. Spirit was now a stationary platform.

With the Martian winter approaching, the operators positioned the solar panels as best they could to stay warm through the cold months. On 22 March, Spirit made its last communication with home. NASA kept listening but heard nothing. The mission was officially ended on 25 May 2011, when NASA announced they would no longer attempt to contact the rover. The brave rover had driven over 7.7km, across hills and through some of the most dangerous terrain ever traversed.

Now, attention turned to Opportunity. The rover was midway through its marathon drive towards the Endeavour Crater. It reached the crater in 2011 and spent the next seven years skirting along Endeavour's rim.

But in June 2018, the engineers noticed a dramatic drop in the solar panels' power generation. A huge dust storm was forming on Mars. Dust storms are common on the Red Planet, but every few Martian years there's one big enough to encompass the entire planet, blanketing the skies and blocking out the Sun from any unfortunate solar panel. Opportunity was right underneath the point where this one was starting.

On 10 June 2018, Opportunity sent its last message, measuring a sky opacity higher than had ever been seen on Mars before. This last stream

Opportunity looks back over its tracks on 4 August 2010. (NASA/JPL–Caltech: www.jpl.nasa.gov/spaceimages/details.php?id=PIA22928)

of data was poetically summed up by science reporter Jason Margolis on Twitter, 'My battery is low and it's getting dark'. These paraphrased 'final words' of Opportunity quickly went viral, once again demonstrating the powerful personal connection the general public had with these two cartoonishly cute rovers.

NASA waited for the storm to pass, hoping the rover would come back once the dust had settled out of the atmosphere. After months of listening they heard nothing. They waited until November, when the region was due a windy period that could clear the solar panels. Still, they heard nothing. On 13 February 2019, NASA made its last call to the rover, accompanied by the sound of Billie Holiday singing 'I'll Be Seeing You'. The rover had lasted 5,111 sols and travelled over 45km. The wind might have given the rovers a reprieve from the dust on their solar panels, but the dust got them in the end all the same.

With over a decade of discovery, it would be impossible to go through all the rovers' achievements here. What follows is a quick overview of the mission's main scientific aims: find out if life could have existed on Mars; characterise the planet's past and present climate; investigate the geology; prepare the way for human exploration.

The first of these goals – determine if there was life on Mars – was never going to receive a definitive answer, at least not from Spirit and Opportunity. The rovers did manage to confirm one of the biggest pieces of the puzzle, though – was there ever liquid water on Mars? The answer was a comprehensive 'Yes'.

There were many geological signs that there might have been water, such as the presence of hematite, clays and carbonates that are often, but not always, formed by liquid water here on Earth. However, the real smoking gun for past flowing water was uncovered by Opportunity when it found a vein of gypsum in the bedrock near Endeavour Crater. The only way such a vein could have been created was if minerals were deposited by water flowing through the rocks.

The rovers were also able to answer the much bigger question of what *kind* of water there was on Mars and whether it might have been able to support life. Initially, the hematite spheres Opportunity found in Eagle Crater suggested a highly acidic climate. While some extremophiles manage to exist in extremely acidic conditions on Earth, they most likely adapted to survive there after growing up in more temperate climes. Whether life could actually *evolve* under such conditions was a different matter.

Meanwhile, Spirit found that the Home Plate region, where it spent much of its time, was created via a hydro-volcanic explosion, where the molten rock came into contact with water. The idea was reaffirmed from an unexpected source – Spirit's limp wheel. As the rover dragged its broken leg across the surface, it dug a trench and the overturned soil was 90 per cent silica. Creating such silica-enriched soil requires hot water. On Earth, that environment is found around hydrothermal vents on the ocean floor. These vents are teeming with life and could have been the place where the first organisms on Earth came into being. Could life have started in such a place on Mars, too?

While conditions were still more acidic than life on Earth prefers, there were some places where things were milder. Spirit found a type of rock called a carbonate. These only form under neutral conditions and dissolve in acidic water, meaning, for this rock at least, there must have been a time when Mars wasn't as deathly acidic as Viking and Opportunity had found. It also suggested there'd been a rich carbon dioxide atmosphere supplying the raw materials needed to make such carbon-rich rock.

The friendliest conditions for Mars life, however, were found by Opportunity at Endeavour Crater in 2013. The rover found a rock named Esperance, which was loaded with clay that only forms in neutral pH water.

It seems that Mars had an era where the water was neutral and possibly supportive to life, but the water later grew bitterly acidic before disappearing entirely. The rocks of the previous era were already buried and protected but were later excavated by impacts bringing them to the surface, where the rovers could take a look.

With these discoveries, Spirit and Opportunity achieved their first three goals: investigate potential for life, the geology and the climate. The last question was harder to answer. What did the MERs do to prepare for human exploration? In terms of telling us how to live or operate on Mars – not much. What they did provide was a reason to continue exploring. There had been water on Mars, and it was water that might have been conducive to life.

The rovers also highlighted the limitations of robotic exploration. It had taken days to perform tests a human geologist could have done in minutes and with a much more limited range of experiments than a crewed mission would have access to. Although the rovers had been a great success, one fact still seemed true – the only way we will ever find life on Mars is if we go over there and have a closer look ourselves.

19

THE PHOENIX RISES

Peter H. Smith was a man whose career had been plagued by bad luck.
By the early 2000s, he was already a veteran of planetary exploration,
having supplied cameras that ended up on the surface of both Mars and
Saturn's icy moon, Titan. Not all of this hardware got there in one piece.
Both Beagle 2 and the Mars Polar Lander had carried Smith's creations
– both failed. Then a third project, the Mars Surveyor 2001 Lander,
never even got the chance to fly as it was cancelled, even though half of
the components had already been built and were currently sitting in the
clean room of Lockheed Martin.

Then in 2001, NASA sent out an announcement of opportunity,
requesting designs for the first mission in a new initiative, the Mars Scout
Programme. Set up to help NASA meet its 'one mission per launch
window' target, the call asked for designs for a spacecraft costing no more
than $325 million, to launch during the next free slot. The 2005 launch
window would be taken up by the Mars Reconnaissance Orbiter (MRO),
which would create high-resolution maps of the planet as well as acting
as a communications hub for future missions. That meant the Scout mission would take up the 2007 window.

At first, Smith tried to ensure his place on a Mars mission by contacting
all of the twenty or so teams and offering to design their cameras. Seven
took him up on the offer. Then he got a call from the Ames Research
Centre, one of NASA's many development laboratories. They had an idea
– bring Mars Surveyor out of storage and build a new mission around it,
using technology developed for the Mars Polar Lander. And they wanted
Smith to be the principal investigator. Although his many failures left him

wary of the responsibility, Smith accepted. And so, from the ashes of two failed missions, a new one arose: Phoenix.

Phoenix was designed backwards. While most missions are built from the ground up to perform a specific task, when Smith asked what Phoenix's purpose would be, he was told, 'That's your problem, you're the principal investigator.'

Luckily, Smith didn't need to look far for inspiration. In spring 2002, the Odyssey Orbiter, which had been at Mars since 2001, discovered large concentrations of hydrogen at the Martian poles. Hydrogen makes up two-thirds of a water molecule, so high levels of the gas usually means that all-important substance isn't too far away. The Odyssey data seemed to suggest there was ice hiding just underneath the surface around Mars's frigid poles.

Reiterating Spirit and Opportunity's mantra of 'follow the water', Phoenix was designed to dig down to find this ice. A stationary lander might seem like a step backwards after two hugely successful rovers, but it meant that the probe could take its time and really entrench itself without having to worry about picking itself up and moving on again.

In many ways, Phoenix was similar to the Viking missions. As well as looking for water ice, the lander would investigate its surroundings for the chemical ingredients of life. It would also investigate potential hazards to emerging lifeforms, such as high salt levels. To search for these signs, the lander had a robot arm. Although this was based on the same design as Spirit and Opportunity, it could reach almost twice as far and dig down 50cm. It would also have Viking-like ovens to cook soil samples, releasing their gases so that spectrometers could analyse their elemental make-up. Phoenix's ovens would reach a scorching 1,000°C, hot enough to start breaking down the more complex organic molecules that could indicate life. The lander would also carry a microscope, as well as a meteorological station to keep an eye on the weather.

All of this would be strapped onto a circular instrument platform just over 1.5m wide. This was perched on three legs – the Surveyor landers had proved the configuration worked, and no one had uncovered a better design for balancing on uneven ground. Jutting out from either side would be two hexagonal solar panels, like a set of giant Micky Mouse ears. Once fully deployed, the lander would be around 5.5m long – about the size of a large car.

As it was made with already developed and tested technology, Phoenix would be low cost, low risk and able to investigate an area of Mars that no other lander had been to before. It was a winning combination and it was selected to go to Mars.

Building Phoenix, however, proved to be a bit more complicated. The initial plan had been to get the Mars Surveyor Lander out of storage, strap some new hardware to it and send it to Mars. But, as they began to take stock of what the lander was like, the Phoenix team began to realise there were pieces missing, scavenged by some other team before them. Being unsure exactly what was on the lander – and what state it was in – the quality assurance team insisted that every circuit and joint was checked.

As the Phoenix team looked at each element of the lander, they began to find potentially fatal flaws in the design, which needed to be fixed. For instance, the parachute cables didn't come loose at Martian temperatures, so the Phoenix team added a heater. In retrospect, finding such problems was unsurprising – Mars Surveyor had been based on the design of the Mars Polar Lander, which had itself crashed – but with each flaw found and fixed, the cost of Phoenix began to rise, and it soon went over budget.

The mission was almost cancelled but the agency was determined to send a mission every twenty-six months, and so Phoenix cheated death. Its eventual price tag was a still-affordable $386 million. The spacecraft launched towards Mars on 4 August 2007, arriving at the planet ready to make its landing on 25 May 2008.

For the first time, the landing would be watched by three of the orbiters already at Mars – MRO, Odyssey and Mars Express – which had reoriented themselves to watch the spacecraft making its 7-minute descent.* They were meant as a failsafe in case the lander crashed, so at least they would have some record of what happened. After all, rather than using the airbag system that had served the previous three Mars landings so well, Phoenix was using thrusters. Mars Polar Lander had used thrusters, and that had crashed. Would Phoenix suffer the same fate?

Smith was at the Operations Centre on the day of the landing:

The press officer had written press documents for every kind of failure in case he had to explain it to the public. As we got past each one of

---

*   This descent, which is the same length of time for almost every Mars lander, is often
        referred to as the 'seven minutes of terror', owing to the fact that all the engineers
        can do while the spacecraft they've poured their heart and soul into hurtles towards
        Mars is watch and worry.

those milestones that he had prepared a failure mode for, he was tearing them up and throwing them in the air like confetti.

As the lander finally approached the surface, the announcer began to count down the distance to the surface, with the room chanting along, until finally … 'Touchdown. We have signal.'

Two hours later, the first images of the landing site – a place called the Green Valley in the Vastitas Borealis in the planet's north polar region – began to come down. The site had been picked to be a 'mattress site' – soft and safe to land on. Like the terrain Opportunity had come down on, the landscape was smooth, bar the odd pebble – no boulders, no craters, no distant mountains. As Smith put it during a press briefing after the landing, 'I know it looks a little like a parking lot, but that's a safe place to land. That makes it exactly where we want to be. Underneath this surface, I guarantee there's ice.'

There was, however, one strange feature they'd never seen on Mars before. The ground was broken up into polygons, each around 5m across and separated by bright troughs. The Antarctic has similar features, caused by the constant freezing and thawing of the ground, an explanation that worked just as well for the Martian pole. As fortune would have it, the lander had come down close to the boundary of one of these, allowing it to investigate two of the polygons (named Humpty Dumpty and Cheshire Cat) and the trough (Sleepy Hollow) between them.

An even bigger surprise came when the robotic arm took a photograph underneath the lander to check there'd been no damage when it touched down. Right under the lander was a bright smooth material covering the surface. It was ice! Or at least, the dirty icy-soil mixture they'd hoped to find. The landing thrusters had blown away the upper layers of dust, exposing the ice. They'd been on the planet just a few sols and had already found exactly what they'd come to look for.*

The robotic arm made its first contact with Martian soil on Sol 7, and soon got to digging. The first dig site, called Dodo, was a shallow 3cm deep but even this was enough to expose a swathe of white material. While this could have been salt or silica, it disappeared after several days, suggesting it was the ice they were looking for.

---

*   The image was later called 'Holy Cow', after someone in the audience made just that exclamation when it was displayed for the first time.

Bright patches of ice revealed by the Phoenix lander's thrusters. The photo is known as the 'Holy Cow' image after a researcher made the exclamation during the meeting when it was revealed. (NASA/JPL–Caltech/University of Arizona/Max Planck Institute: www.jpl.nasa.gov/spaceimages/details.php?id=PIA11379)

On Sol 12, the ovens baked their first sample. One of the oven doors was stuck half open, but Phoenix kept dumping in material until there was enough to analyse. After running through the data, NASA finally made the announcement they'd been waiting years for on 31 July 2008 – Phoenix had made the first direct detection of water on the surface of Mars.

This didn't mean that all conditions were rosy for Martian life. In August 2008, NASA announced that the lander had found evidence of perchlorates. These chlorine-bearing chemicals act as an antifreeze, keeping water liquid. This meant it could be possible to have liquid water at temperatures that were much lower than had been anticipated.

Unfortunately, such water would be extremely chlorinated. Seeing as, on Earth, we chlorinate pools to kill bacteria, it was another check against there being life on Mars. The find also shed new light on Viking, because if there had been perchlorates in the samples it tested, its life-finding experiments would have almost certainly failed to find any microorganisms, even if they were there to find.

By the end of August, Phoenix had reached the end of its ninety-day planned mission but was still going strong. During that time it discovered magnesium, sodium, potassium and calcium – all elements you'll find on the side of a bottle of multi-vitamins. Even if current life wasn't exploiting them, any future Martian colonists would be grateful to have them in the soil.

Phoenix lasted another month and a half, but the Martian winter was rolling in and the light was failing. The lander wasn't designed to survive the harsh polar winter and the solar panels soon struggled to generate power. On 2 November, Phoenix stopped communicating with Earth. With no power, the heaters on the lander couldn't keep the carbon dioxide from condensing onto its solar panels. They weren't designed to hold the extra weight and almost certainly buckled. Although NASA attempted to re-establish contact in 2010, the ice seemed to have extinguished Phoenix's flames – it was gone.

Despite its short run, Phoenix gave a definitive answer to the biggest question of Mars: it had discovered unequivocal evidence of water – not in Mars's past, but in the present day. It was frozen into the soil and useless to life, but it was still water.

20

A CURIOUS ROVER

With Phoenix done, the decade-long plan set down after Pathfinder was coming to its end. NASA had followed the water and shown that Mars was once much warmer and wetter than it is today. This water might have been toxic to Earth life, but that didn't mean life with a different set of survival skills couldn't have evolved on the Red Planet. Nor did it preclude the possibility of human life walking on the surface in the future.

Life on Mars. It was an investigation that NASA had been skirting around for decades, but it was still a goal the agency desperately wanted to achieve. With seven successful landing missions under their belt, they now had the tools for a mission that could tackle the question more directly.

Realistically, the only way that life will ever be discovered on Mars (if it even exists) is if human scientists can get a look close up. One way to do this is a crewed Mars mission. NASA hadn't forgotten its dreams of landing a human on Mars. Although many people at NASA and beyond have mulled over the idea for years, the agency had never sat down and realistically considered from end to end how such a mission could be done. In December 2004, NASA ran the first of many workshops that brought together the EDL teams from the Mars missions and the engineers who had built the Space Shuttles and the International Space Station.

Together, they hashed out the requirements for such a mission, with one team often pointing out the flaws in another's idea. For instance, at one point Jack Schmitt, who had walked on the Moon during Apollo 17 and gave an astronaut's perspective on the matter, offhandedly mentioned that any human mission would need the ability to abort the landing at any time. The news shocked the EDL teams. They didn't have a clue how you could even begin to approach that task, let alone do it reliably.

One thing became obvious – NASA was woefully unprepared to launch a human mission anytime soon. But perhaps if they couldn't yet bring the human to Mars, they could bring Mars to the humans and stage a sample-return mission.

Whether it was a human or sample-return mission, finding life on Mars was going to require some serious machinery on the surface that was far heavier than what they could land at that time. The MERs had pushed the airbag landing system to its limits, while the parachute-and-retrorocket technique was still limited to landing on relatively flat, boulder-free areas. If they were going find life, then the mission needed to be able to reach the areas where life was likely to have evolved – preferably without having to spend years driving over to it. NASA needed a way to accurately land a heavy payload on the surface.

Thankfully, a working group had been considering just such a mission – the Mars Smart Lander. Initially, the mission was meant as an investigation into landing something large on Mars. As time progressed, the token scientific payload grew into a fully fledged rover, five times heavier than anything the agency had sent before. The project, now one of NASA's flagship missions, was renamed the Mars Science Laboratory, helpfully maintaining the same acronym – MSL.

As well as testing the technology that would be needed to perform a future sample return – and a human mission further down the line – the rover's scientific aim was to examine the Martian environment's biological potential. Again, the focus would be on water and cataloguing the chemical resources that microbes might use. The rover would also look at ways in which the planet might be hostile not just to current life but future human settlers. As Mars has neither a magnetic field nor much of an atmosphere, radiation from space blasts the surface and the MSL rover would investigate just how inhospitable this made the surface.

To do this, it would have a mammoth ten instruments. Once again, there would be a robot arm, similar in design to those found on Spirit and Opportunity, an APXS spectrometer, a stereoscopic camera and a meteorological station. However, the rover's new headline instrument would be the Sample Analysis at Mars (SAM) experiment, a soil analysis machine capable of hunting down organic compounds.

One big problem, as ever, was power. If the rover used solar panels, it would take not just winglets but full-blown wings just to keep the motor running. Instead, it would use a radio thermal generator (RTG) like those used on Viking. However, even one of these would produce only 110W of power. Running at full capacity, the rover would need 1,500W.

Fortunately, RTGs run for 24 hours and 34 minutes per sol. When the rover shut down for the night, the batteries could recharge to give the extra oomph needed during the day. It also had the bonus of acting as a heater, keeping the body of the rover warm and cutting out the need for power-hungry heaters.

While the designers were working out the details of what was going on the rover, there was one problem still needing a solution – how to land MSL on Mars. So far, every mission had at least partly relied on parachutes to slow down, so they seemed a good place to start. But it wasn't a simple case of scaling up the design to cope with the extra weight. A larger chute takes longer to open. Too big, and the lander will hit the surface before the parachute had time to deploy.

Parachutes are also very imprecise. While the EDL team could hit a general area using parachutes, the potential landing zone could be several hundred kilometres long. If NASA wanted to target specific landmarks rather than a general region, they needed something with more control.

The solution eventually came from an audacious plan: land the rover directly on its wheels rather than encasing it in a protective landing plat-form. The rover would be suspended beneath a rig of jets. However, instead of just being there to take the worst of the velocity away, these would be able to direct the landing rover with extreme precision to a landing zone just a few square kilometres in size.

It was a crazy plan but as NASA Administrator at the time, Mike Griffin – who had an unusually keen grasp of the technicalities of spaceflight for upper management – said, when he first heard the idea, 'It's the right kind of crazy. So crazy it might just work.'*

With the new landing system, named Skycrane,** in the works, the project was given the go-ahead in 2005 for a launch in 2009. While the MSL team were ecstatic at the news, the feeling was not universal across NASA. In April 2007, Alan Stern was made the Associate Administrator for the Science Mission Directorate, the most senior scientific role within NASA.

---

*   R. Manning and W.L. Simon, *Mars Rover Curiosity: An Inside Account from Curiosity's Chief Engineer* (Smithsonian Books, 2014) p.70.

**   Named after the Sikorsky S-64 Skycrane heavy-lift helicopter.

The science wing of NASA had once again grown used to ballooning budgets, and Stern decided it was time to start taking a fiscal hard line.

Although many missions were reviewed, it was the Martian programme that bore the brunt of his ire. Stern's background was in the outer planets, and he was the principal investigator for the New Horizons mission that would soon begin its way to Pluto. He was a good scientist, but Stern had little aptitude for diplomacy. He felt there was an undue emphasis on the Red Planet and made his stance publicly known, declaring it was, after all, 'just another planet'. While many were happy to see less-explored worlds given their due, he was seen as gutting the high-profile, media-friendly Mars missions to do so.

Stern wanted the Mars missions to actively work towards a sample return, rather than some vague pledge for the future. But he wasn't giving them any more money to do so, meaning the Mars programme was going to have to abandon its quest to launch one mission in every launch window.

Initially, Stern wanted to push back MSL's launch date to 2011, reorienting it as a precursor mission to a sample return – for instance, creating sample caches that a future mission could pick up. However, during his tenure as administrator, Griffin had been slowly pushing back launching a sample-return mission to Mars and was riled at the interference.

Stern presided over MSL's critical design review in June 2007, where the entire project was grilled by a panel of independent experts to ensure it was on track and feasible to continue. The project, so far, had been given the sizeable budget of $1.6 billion. A third of this had already been spent. Although this allowed for some overruns, it was becoming apparent that the budget had been underestimated by as much as 15 per cent. MSL requested more money – Stern refused. Their budget was the budget. The MSL team would have to descope the project and start removing instruments.

To those who had been working on MSL, the news was devastating. The instruments had all been carefully chosen to work together. Removing any one would seriously impair the scientific capability of the entire rover.

The final blow came over a relatively small amount of money – $4 million. Stern was unwilling to make other areas of planetary exploration suffer to free up money for MSL, firmly believing that 'Mars should pay for Mars'. However, he had no issue with taking the money away from other Martian missions. He found he could easily cut $4 million from the budget set aside for the Spirit and Opportunity rovers. By this point in 2008, their cost-to-scientific-return ratio was beginning to dwindle. However, the cut would mean putting one of the rovers into hibernation and limiting the other.

Stern made the mistake of discussing this idea publicly before bringing it to Griffin. It soon leaked to the press. The public was outraged at the idea of 'killing off' one of the charismatic rovers. Griffin publicly stated that Spirit and 'Oppy' were safe from the chopping block. Stern, feeling Griffin lacked faith in his decisions, resigned.*

With Stern gone, the budgetary controls relaxed, but money was still worryingly tight. Power continued to be a problem. SAM, arguably the lead instrument on the rover, was four times over its allocated power budget – 1,200 watt-hours. MSL needed an expensive, custom-made battery with double the capacity of the first one.

Throughout this turmoil, Rob Manning had been the chief engineer coaxing the project through. Upon stopping to take stock of the situation, he realised with growing horror that they were running out of something far more precious than money or power – time.

In retrospect, Manning realised, a 2009 launch had always been laughably ambitious. There was just too much new technology. But he'd let himself be swept up in the excitement of planning a new mission and promised more than he could deliver. They needed to shift the launch to 2011. 'There's just way too much to do to be ready to launch in 2009, and if we did, I'd be terrified. I wouldn't trust the vehicle,' Manning told NASA management as he begged for the extension.**

Management, however, refused to budge. NASA always pulled it out of the bag at the last minute; they'd do it again. It wasn't until November 2008, a few months before the proposed launch window, that Manning could prove that there just wasn't enough time in the schedule to get to the pad in 2009. There was no choice; they had to delay.

No Mars mission had missed its launch so close the window. Annoyingly, the rover only required a few extra month's work but would have to wait twenty-six months for its next chance in 2011. With two years' worth of

---

*   Stern doesn't come off very well in this story when you only hear about it from the Martian side. In truth, his argument that the rest of the solar system was being abandoned in favour of Mars was entirely reasonable. In many ways, he was simply trying to revitalise NASA's Martian exploration, but unfortunately, he didn't always do so in the most diplomatic of ways.

**   Manning and Simon, *Mars Rover Curiosity*, p.113.

extra man hours, the delay would be costly. The project was far too high profile to mothball and Congress agreed to give MSL an extra $400 million funding – but it would start in 2010. With only a few weeks' notice, the assembly team had to finish off what they could and prepared to store the rover for the next year.

The project constantly made headlines as the media, opposing scientists and politicians alike questioned the cost and management of the mission. However, there was one ray of hope in the gloom. The mission had had its naming essay contest and the winner was 12-year-old Clara Ma from Kansas. Her essay was a testament to one mankind's best qualities – Curiosity.

As the hiatus began, many of the MSL engineers went off to work on other projects. Meanwhile, Manning began to create an epic to-do list, known as the Manning List, detailing every issue on the rover that needed resolving. The items on his list varied from simple tasks, such as building a wind guard to stop samples blowing away, to working out how to test the Skycrane while remaining on Earth.

By the end of 2009, most of the items on the Manning List had a solution; they just needed to be implemented. As the engineers and fabricators returned, the rover was taken out of storage. Manning ordered his workers to make their way through his list. Money was still tight and so a new policy took shape: only fix bugs that will seriously harm the mission. If there was a more minor problem, then they would have to learn to live with it.

By 22 July 2010, the rover was assembled. And it was huge. Whereas the MERs had been shorter than the average person, Curiosity was the size of a truck and weighed 900kg. Although it was similar in a grand scale to the design of Spirit and Opportunity its appearance was far more industrial. Without the solar panels creating a sleek finish to its top, Curiosity's beams and cabling were exposed for all to see. Its struts and robotic arm were bulkier in order to deal with its extra weight. Even its central body was stripped of the gold foil that had kept the previous rovers warm. The MERs had been slender show ponies; this was a cart horse.

Despite the extension, testing on Curiosity went right up to the wire. In February 2011, 1,200 problems and faults were still needing to be remedied. The time buffer within the schedule was rapidly diminishing, just as it had done in 2009. It would be a much harder sell to get funding if they needed to delay a second time.

However, the MSL crew *did* manage to pull it out of the bag. On 12 May 2011, a truck rolled up outside JPL ready to carry the rover over

to Cape Canaveral on the Atlantic Coast of Florida.* JPL carried on working on a replica rover to work out Curiosity's remaining niggles.

Alas, one niggle turned into a major problem – the test rover's drill created a short that almost killed the whole rover. They couldn't risk the same
thing happening on Mars and so a bit of last-minute surgery was needed.
No one was particularly enthusiastic about this as a similar eleventh-hour
fix had almost killed MER missions, but there was no choice. With great
care, the team opened up Curiosity's belly pan and made the fix.

A second problem wasn't so easily remedied. One of the test rover's
drills was contaminated with an oily residue. If a similar residue was on
the Mars rover, it could contaminate results. They checked Curiosity's drill
bits and found them clear but didn't sterilise them again after the testing.
The drill bits are one of the few parts of the rover, other than the wheels,
that touch Martian soil. As such, they were meant to be subjected to strict
planetary-protection protocols.

Catharine Conley, NASA's Planetary Protection Officer, deemed that
the rover could still fly to the Red Planet, but the drill was not allowed
to be used in any area where a living cell might be able to survive,
i.e. anywhere there was water or ice. Fortunately, Curiosity was
intending to search for past water, not present, and was heading towards
the dry equator.

Despite all these problems, Curiosity made its way to the launch pad on
26 November 2011 and headed towards Mars. Its destination was the
Gale Crater.

The location was picked after several years of vigorous debate,** as the
Skycrane's precision meant that the rover could be delivered within a few
kilometres of the most interesting features on Mars. NASA was once again
following the water, as there were signs that the bottom of Gale Crater's
basin had once been home to a large lake. However, the real selling point
for Gale Crater was Mount Sharp – a 5.5km high peak in the crater's
centre. The hope was that the mountain would have several exposed

---

*   In a musical interlude, Manning accompanied its departure by playing 'When the
     Saints Go Marching In' on the trumpet.
**   In all areas of academia, 'vigorous debate' can cover everything from polite but
     emphatic discussion to full-blown fist fights.

layers. As the rover drove up the mountain, it would be able to inspect these layers, revealing Mars's geological history right back to its warm and wet past.

The rover began its descent towards Gale Crater on 5 August 2012. The mission had, by this point, garnered significant public interest and over 3 million people tuned into watch the live stream as NASA performed the first ever full run of a brand-new landing system, setting down a flagship mission with a seven-figure price tag. No pressure then.

In the Control Room at JPL, the EDL team had gathered. Amongst them was systems engineer, Bobak Ferdowsi. Soon, his face would be on hundreds of internet memes, commenting on his striking mohican hair-cut, dyed to look like the American Flag for the special occasion. In this moment, however, he was simply there to turn off the uplink signal to the spacecraft. From now on, the lander was on its own.

It would take the rover just 7 minutes to reach the surface, a time that had come to be known as the 7 minutes of terror. The lander hit the atmosphere at 21,000km/h. The extra weight of the lander meant the heat shield had to be coated in a new material that could bleed away the heat, but which had never been tested on Mars. It did its job first time, though, and soon the lander was travelling a positively sedate 1,600km/h. Then the 21.5m-wide parachute deployed, slowing the falling rover down to just 320km/h.

At 4 minutes and 39 seconds into the descent, the heat shield jettisoned to reveal the radar system that would guide Curiosity to its landing zone. Just over a minute later, the parachute cut away. The thrusters took over, dodging the parachute and back shell as they fell around the rover, before guiding the rover in closer to its target.

Just 23m from the surface and now travelling at 3km/h, the rover itself dropped below the Skycrane, suspended by three cables as it fell the final few feet towards the surface. A few moments later, its wheels touched Martian soil. The cables immediately cut away and the Skycrane jetted off to safely crash several metres away.

Curiosity had landed. The Skycrane had worked.

Curiosity's first look at Mars was through its dust-flecked lens cap but the image was clear enough to see a wheel resting on the ground. Beyond these wheels was an empty-looking landscape covered in gravel, with the dust blown away by the wind. Fortunately, this rather featureless area wasn't the reason that Curiosity had come to Gale Crater. Its destination was Mount Sharp, 6km away.

Before Curiosity could start the hike, it needed a personality transplant. It would never have to land on Mars again and so the EDL program was deleted from its hard drive, making room for the 3.5 million lines of code that would operate the rover on the surface. Merely transmitting the code took four days.

A week later, the rover woke up a new woman. It gave itself a quick once over to make sure that everything was in one piece and found, alas, it had one casualty of the landing – a wind sensor. It seemed no one had told the Spanish team who created it that their instrument might get pelted with debris during landing. A pebble had hit one of the sensors and destroyed it. Thankfully, a second one was still working.

Two weeks after landing, the rover stretched its legs for its first ever real drive. Unlike Spirit and Opportunity, Curiosity was an independent rover and didn't need its hand held through the procedure. Artificial intelligence had come on in leaps since the MERs had been developed, and with it, Curiosity's autonomous navigation. The rover was able to take images of its surrounding, analyse them and decide on the best way to avoid boulders and craters, all without any input from Earth. The driving team could lay out a general path they wanted the rover to take and Curiosity would step in to take care of the rest.

While the rover had been sitting on the surface, the many spacecraft in orbit scouted Curiosity's landing site. Although the rover was there to study Mount Sharp, there was always the risk that it would conk out before ever reaching its target, and so the operations team planned out several science stops along the way to ensure it completed at least some of its objectives.

Its first spot was an area three to four weeks' drive away named Glenelg, where three different types of terrain converged.

Before heading off, Curiosity stopped to take one last self-portrait at its landing spot, now named Bradbury Landing after the recently deceased sci-fi author, whose *Martian Chronicles* short stories had helped to shape the public's idea of Mars in the 1940s and '50s. Taking a selfie is a bit more of a palaver for a rover than it is for you to snap one on your phone. The image had to be stitched together from dozens of photos taken by a camera on the end of the robotic arm.*

---

*   Spirit and Opportunity took similar selfies, often featuring as part of conspiracy theories wondering how the rovers were able to take pictures of themselves. If you look closely at the selfies you can often see the artefacts where the images are stitched together, most notably around the end of the robot arm, which vanishes halfway down, at the point where the camera can't see it anymore.

The image was created for the operations team to check everything was where it should be, but quickly became a hit when it was shared on social media. The rover was proving to be as popular as its predecessors, quickly racking up over a million followers on Twitter and drawing messages from everyone from Richard Branson to Britney Spears.

Curiosity reached Glenelg on Sol 52. Over the next year, the rover lumbered towards Mount Sharp, ultimately travelling 8km as it made detours around craters and visited places of particular scientific interest. Eventually, it arrived at its destination in September 2014, ready for the real work to begin. The rover readied itself and began the long, difficult climb up the mountain. As it wound its way towards the peak, it took several pit stops to take a look around and examine the many layers of Mount Sharp.

MRO images showed there were several different 'units' up the side of Mount Sharp. At the base were the Bagnold Dunes – a series of rippling sand hills – of which Curiosity skirted the edge. From there, the rover drove into a region of sedimentary rocks known as the Murray Formation, a clay-rich region created at a time when Gale Crater was filled by an enormous lake.

Next up was the hematite unit, then the clay unit and finally the sulphur unit, so called because they appear to be rich in these various minerals. As of writing, Curiosity is in the clay unit, having traversed over 21km.

Today, the rover is beginning to show her age. The current biggest concern is its wheels. These are made of a ring of material with several metal ridges to provide grip.* The fabric of the wheels began to develop holes soon after landing on Mars. This wasn't too much of a problem at first, but controllers had no idea if the holes would grow bigger, if the wheels would become unusable, or if everything would be hunky-dory. Luckily for NASA, it seemed like the latter. Although the wheels would be a recurring thorn in their side, it was one they could learn to deal with, and Curiosity has been making progress despite her aching feet.

A slightly more worrying issue was when the rover suffered a major computer fault. Curiosity switched to its redundant computer (proving why redundant systems are always worthwhile) and operated on that for

---

*   When JPL showed off an early prototype of Curiosity to the world, it had a large JPL sticker on the side. Wanting to appear like a unified agency, NASA banned JPL from putting their logo on the final rover. They retaliated by arranging the metal tyre treads to spell out 'JPL' in Morse code. With every metre Curiosity travels, it etches the name of its creators onto the surface of Mars.

many years. However, in 2018, shortly after the dust storm that killed Opportunity, the back-up computer began to develop a fault too. For a while, many feared Curiosity might be too scatter-brained to go on, leaving Mars without a rover for the first time in fifteen years. Luckily, the issue with the primary computer had been fixed, and Curiosity switched back to using that.

Despite these issues, Curiosity has made some epic discoveries – and continues to do so. As with Spirit and Opportunity, to list everything that it has done over its seven years on Mars would take up a book on its own and much of its work is still ongoing. Instead, here is a list of some of the rover's greatest hits.

By the time Curiosity came along, there seemed little debate over whether Mars had once had liquid water flowing across its surface. Curiosity backed up this idea by showing that the Gale Crater seems to have been the bottom of a now-evaporated lake, lined with beds of clay and sedimentary rocks and peppered with rocks that have been smoothed by an ancient river flow.

The real question now was what this water was like, how long it was on the planet and why it disappeared.

Unlike other areas of Mars, the water of Gale Crater's past lake appears to have been drinkable, meaning it could have hosted life. This depends on (amongst other things) how long the lake stuck around for. While there was enough water on Mars to create mass floods and fill craters, there's no indication how long the water remained on the surface. It could be that water ebbed away across the planet over several hundred million years. Or, it could be that for most of that time the planet was a snowball, with all or most of its water frozen, but then the planet briefly (geologically speaking) warmed up – perhaps due to a volcanic eruption or a meteor impact – causing the ice to melt and form lakes and rivers. Regions within the clay unit suggest that the water history of Mars might be more complicated than it slowly evaporating away over time, but exactly how complicated is going to take a bit more research.

Water wasn't the only thing Curiosity was looking for. When the rover examined the rocks, it found several other key ingredients for life such as sulphur, nitrogen, oxygen, phosphorus and carbon, as well as more complex organic chemicals. This latter find was a good omen as Curiosity only looked at the top few centimetres of soil, which is exposed to the Sun. As solar radiation destroys organic chemicals, it's entirely possible that there are even more of these biological building blocks hiding under the surface.

The biggest find in terms of organics, however, wasn't in the rocks but in the air. Throughout its time on Mars, Curiosity has been sniffing the atmosphere to see what gases it contained. There was one particular gas that scientists were on the hunt out for: methane. Previous missions had found hints of it, but Curiosity was built with an instrument, the Tunable Laser Spectrometer (TLS), which had a laser tuned to a wavelength of light that's absorbed by methane. The laser was bounced through the air before being measured by a sensor. If the light level dropped, then methane was present.

For the first year of its stay, Curiosity detected no methane on the planet, but kept sniffing, just in case. On Sol 466, during the planet's warm summer, it smelled a hint of the gas. A few sols later, they tried again and sure enough, there was a clear signal for methane. It was only on a level of around seven parts per billion (ppb) by volume (to put that into perspective, Earth's current level of methane is around 1,800 ppb), but it was still there. The puff of methane lasted for approximately 60 sols, then disappeared.

A Martian year later, when summer rolled around again, the rover smelled another peak in methane, suggesting this was a seasonal occurrence. What's causing it, though, is a matter of much debate. Methane is, on Earth, often produced by life, so it's not surprising that one of the most alluring ideas is that there's a population of microbes in Gale Crater that wake up during the summer months, belch out methane for a few sols then go back to sleep. However, there are more than a few geological processes that could spew methane into the air without the need for microbes and a number of these might also be triggered by seasonal warming.

In 2016, ESA sent the Trace Gas Orbiter to Mars. As the name suggests, the orbiter was looking for traces of gases in the Martian atmosphere, most notably methane. But in its first two years, it hadn't found any of the gas on a global scale. This doesn't mean the Curiosity results are wrong – just that these puffs are small and confined to a local area.

Curiosity has shown that if there was life on Mars in the past it would have had water to drink, the raw chemical ingredients it needed to eat, and an atmosphere to breathe. Why this, by no means, makes Martian life a certainty, it is a definite case for it being a possibility.

The rover also felt out signs that the planet's atmosphere may have been thicker in the past. The rover detected the atmosphere has more of a heavy variety of carbon than it's expected to have had when it formed. As lighter atoms escape a planet's atmosphere more readily than heavier ones, this mismatch suggests some of the original atmosphere must have escaped. This finding was later backed up by the MAVEN Orbiter, which arrived at the planet in 2014 and found that the upper levels of the atmosphere were being lost to space.

However, Curiosity wasn't just looking at whether life exists on Mars now, but whether it might in the future. The Skycrane system proved NASA could put down heavy loads on the surface with a great degree of accuracy, a skill that would be vital if humanity ever wanted to set up a long-term base on the surface.

It did show that there might be some problems in keeping them alive during the mission, however. During the journey, Curiosity measured 1.8 millisieverts of radiation a day – the equivalent of getting a full-body CT scan every five or six days. This would lead to astronauts having a 5 per cent increased risk of getting fatal cancer over the course of the round trip to Mars, which is higher than NASA's acceptable limit of 3 per cent. If humanity plans on regularly travelling to Mars, the radiation issue is one that will have to be addressed and future mission planners will use Curiosity's data to do so.

Curiosity was still running as of the end of 2019. Provided its computer keeps behaving itself, there is no reason Curiosity shouldn't keep going for several years more. There's no telling what great discoveries the rover still has left to make.

21

AN INSIGHT INSIDE OF MARS

After the success of Curiosity's landing in 2012, the world was looking to see what NASA would do next at Mars. Even as the rover was touching down to examine the Martian surface, a new group were working out how to take a look inside the planet itself.

While initially NASA had been concerned with the physical planet itself, the Allan Hills meteorite had changed their emphasis to hunting for signs of life and paving the way for human exploration. Any geological investigations there had been mostly focused on the role of water on the planet and its past climate.

What the landers had failed to examine was how the planet had formed, evolved and grown into what we see today. Mars's geology is frozen in its early evolution and hasn't been changed by the tectonic and volcanic activity that has altered Earth over the last few billion years. By studying Mars, we not only learn about the planet itself, but about what all terrestrial worlds, including our own, were like during the early eras of their lives.

That isn't to say that people hadn't been trying to get a geophysical lander on the flight list in the intervening years. JPL's Bruce Banerdt had been involved with several projects aiming to send a geological mission to Mars. Most of these had been network missions, with multiple landers keeping a finger on Mars's pulse – its seismic activity – but most of them met with little success. They were too complex; too expensive; too boring.

As he was determined to get a geological lander on Mars, Banerdt began looking at what seismic investigation you could achieve with a single lander. Conventional wisdom is that you need at least three stations to triangulate the origin of a seismic wave, but a single lander would still be able to tell you a lot about the internal structure of the planet.

In time, he joined up with the French team who built the seismometer for the Russian Mars 96 mission. Having watched that mission fall into the ocean shortly after launch, they were still keen to see their instrument fly. The French were happy to build and – more importantly – pay for another seismometer if there was a NASA mission willing to carry it. By stripping back what else came with the lander, and basing it on the Phoenix design, Banerdt came up with a lander that would cost less than \$425 million, meaning it would be eligible for the 2011 round of Discovery-class missions.

By this point, the Mars Scout Programme, which funded the original Phoenix mission, had been cancelled. Instead, Mars missions were now allowed to compete to become a Discovery programme, an initiative that had previously banned them to give the rest of the solar system a fair shot. 'Most people thought this was a fool's errand,' says Banerdt. 'There's no way you could get a Mars mission selected in the Discovery programme.'

That was almost true. The mission was up against another that aimed to float on the methane seas of Saturn's moon Titan and one that aimed to hop across a comet's surface. Both were much more eye-catching than a mission that would sit quietly on Mars hunting out marsquakes.

Both of these other two missions relied on a new power generator called the Advanced Stirling Radioisotope Generator, and problems with its development meant that it wouldn't be ready in time for either mission to use. They were taken off the table, and the Mars lander got the spot – to the slight consternation of those planetary scientists who believed Mars was getting more than its fair due.

After a year of development, the mission acquired its final name – the Interior Exploration using Seismic Investigations, Geodesy and Heat Transport, or InSight for short. The name reflected the lander's three main goals – measuring marsquakes, gauging how 'sloshy' its interior is, and seeing how fast it is losing heat.

The first of these would be tackled by the Seismic Experiment for Interior Structure (SEIS) instrument, a joint European effort between the French and British. Although Mars is considered a dead world, geologically speaking, there is a question of exactly *how* dead it is. The planet doesn't have the same active geology and plate tectonics found on Earth, but could Mars still have some kind of seismic activity? We know that the Moon – which is a *very* dead world – still has 'moonquakes', from experiments left behind by the Apollo crews.* If the moon is shaking, it seemed likely that Mars might be as well.

These marsquakes, if there were any, would also give a way to look into the planet's interior. When a quake happens, the undulating motion of the rock radiates out from the origin of the quake in a seismic wave. The waves change slightly as they pass through different kinds of rock. This leaves a signature in the wave pattern. By picking apart the signals InSight detects, geologists are able to create 3D map of the planet's layers, showing how thick and molten they are.

The Viking landers had tried to detect marsquakes, but only one managed to deploy its seismometer, and even that didn't work properly. The experiment was mounted on Viking itself, meaning that all the seismometer ever saw was the vibrations of the lander and the wind. Any results it did take therefore couldn't be trusted. Lesson learned – InSight would put its seismometer directly on to the Martian surface and then cover it over with a shield to protect it from the wind and changing temperatures.

These marsquakes, if they exist, could be caused by anything from meteor impacts to residual volcanic activity or the thermal shifting of the planet as it cools. Which brings us onto the Heat Transport part of InSight. The Heat Flow and Physical Properties Package (HP3, pronounced 'HP-cubed'), designed by the German space agency, aimed to measure the flow of heat into and out of the planet, helping to understand how Mars cools, both now and in the past.

HP3 was a self-hammering nail, known as the 'mole', that could drive itself 5m down into the rock, trailing a string of temperature sensors behind it. These would be able to measure the subtle changes in temperature as heat flowed outward from the planet's core.

Finally, the lander would measure the planet's wobble (referenced in the 'geodesy' part of the name) using the Rotation and Interior Structure Experiment (RISE). Most planets wobble a bit on their axis as they spin, an effect known as precession. How much depends on what's going on

---

*   Moonquakes are much weaker than earthquakes, as they're almost always caused by external forces. Some are created by meteor impacts, others by the gravitational pull of the Earth, a reversal of the Moon's effect on the Earth's sea tides. A few are from the rock expanding during the hot day. Finally, there are shallow moonquakes, which occur in the top 20–30km below the surface and can reach up to a 5.5 on the Richter scale. No one really knows what causes these powerful quakes, but they could cause problems for any future lunar settlers.

inside the planet because one with a solid core wobbles a lot less than one that has a liquid core sloshing around all over the place.*

To measure precession, InSight needs a little help from Earth. The spacecraft bounces signals between Mars and Earth, but rather than trying to convey information, the transmission is measuring how fast the lander and Earth are moving in relation to each other. Over time, the team will be able to use this to accurately measure Mars's wobble, then use that to work out how liquid the planet's core is.

All three of these experiments required a fixed base to work from. After the rovers, the static InSight might seem a little unimpressive, but it was this very immobility that would prove its greatest asset. To deploy its instruments, InSight would use a robot arm salvaged from the cancelled Mars Surveyor 2001 Lander, while a camera would hunt out the best place to set them down. Initially, this was going to be black and white, but Banerdt pointed out this would be the first NASA lander without a colour camera since the 1960s and that might look a bit backward. The ploy worked, and administrators found the extra $4 million needed to upgrade them.

As development of SEIS progressed it became apparent that the lander needed a sensor to measure how much the wind shook the seismometer. Fortunately, the Spanish team who had built the wind sensors for Curiosity – the ones destroyed by flying debris during landing – had a couple of flight spares they could donate.

Before deploying its instruments, InSight looked almost identical to Phoenix – a circular instrument platform on tripod legs, flanked by two hexagonal solar panels. But while Phoenix only lasted 157 sols, InSight would be designed to keep going for years. Fortunately, as it would be based near the equator, away from the frigid poles, it wouldn't have the same problems with carbon dioxide ice that Phoenix had suffered during the cold winter.

After two years of development, the lander began construction in May 2014. JPL were once again in charge of overseeing the project, but contracted Lockheed Martin to build the main body of the lander. While this part went to plan, creating the instruments was a more fraught process. Banerdt says:

---

*   If you want to see what I'm talking about, get yourself one raw and one hard-boiled egg. Place them on a table and spin them. You should see that the boiled egg, which is solid inside, spins nice and neatly. Meanwhile, the raw egg wobbles all over the place.

It turns out that when you go from something that works pretty darn well in the laboratory on a clean workbench to a final flight instrument that's radiation tolerant and can survive landing and 100°C temperature variations a lot of things start breaking or not working properly.

While no one issue was major, each problem delayed the whole project by a month or more. Every mission has some contingency built into its construction timeline for just such issues, but by the time the lander was finished in May 2015 this time had almost run out.

Then, during testing the vacuum seal on the SEIS instrument started to leak. The issue was caused by one of the 'feed-through' ports, where the wires run between the enclosed sensors and the outside world. These need to be completely sealed to keep the pressure inside constant, but one was letting air in. The problem was fixed by early October, only for a second leak to appear on another feed-through. Then the instrument sprang a third leak. There was just enough time to fix these, but that was it. InSight could not afford for anything more to go wrong.

By December, the leaks had been sealed and the lander was making its way to the launch site for its final checks before a March 2016 launch date. Unlike all of NASA's previous planetary launches, InSight would take off not from Florida's Cape Canaveral, but from Vandenberg Air Force Base in California. As the launch site was further north it would require a bigger rocket,* but with the rise in private space companies the schedule at Canaveral was getting increasingly congested. By using other launch sites, such as Vandenberg, NASA hoped to free up Florida for the bigger launches that needed to be nearer the equator.

Just three days after arriving at Vandenberg, the testing crew found a fourth leak. It wasn't large, but it would render the lander useless. The problem was fixable, but not in the three months before the launch deadline. InSight would have to wait for the next window in May 2018. Just as with Curiosity, the delay came with a hefty price tag – $150 million. Fortunately (for NASA at least), some of the responsibility for paying the bill fell on the shoulders of the group who made the SEIS instrument, the French Space Agency (CNES).

---

* Rockets get a momentum boost from the Earth's spin during launch. This boost is biggest near the equator, meaning you get more bang for your buck in terms of a rocket's thrust.

InSight returned to storage at Lockheed Martin, while JPL took over developing SEIS's vacuum container with CNES handling its integration and testing. When the lander was taken out of storage in August 2017 and retested, SEIS seemed to be behaving itself.*

By 5 May 2018, all the tests were done. Nothing was leaking and InSight was successfully installed on top of the Atlas V rocket that would send it on its way to Mars. The launch happened at 4 a.m. local time on a foggy night. Onlookers could see little other than an orange bloom streaking off into the western sky on its six-month journey to Mars.

After the complicated landing of Curiosity, InSight's descent to Mars was a return to the simple life, using only a heat shield, parachutes and retro thrusters. It touched down on 26 November.

As InSight wouldn't be studying the local geology, the landing site had been chosen to be as safe (aka – boring) as possible, and the landing team had settled on Elysium Planitia near the planet's equator. Everything about the site – from its topography to its climate – was chosen to be as blankly uniform as possible. Banerdt best summarised the area when he said, 'If it were an ice cream, it would be vanilla'.

The first images from the planet revealed a landscape every bit as flat and featureless as the mission controllers had been hoping. If the team had been aiming for a scoop of vanilla, they'd landed in it.

It took InSight several months to set itself up. Its camera carefully studied its surroundings, allowing the team back at JPL to create a full replica of the landing area. Using a test copy of InSight, they were able to practise deploying the instruments on to the Martian surface using its robotic arm.

On Sol 22, or 19 December to us Earthlings, InSight placed the seismometer on the surface. The arm had to really reach to put the instrument down 1.6m away, as the closer it was the more the electronics and vibrations of the lander would interfere with the seismometer's readings. It was so far, in fact, that the deployment had to take place in the cool of dusk so

---

*   The lander did gain one new addition. NASA ran a 'send your name to Mars' initiative, where members of the public submitted their name to be written on a microchip that was then integrated onto the lander. The first run had gained 850,000 names, but after two years of publicity from celebrities such as the *Star Trek* actor William Shatner, the total list ended up being 2.4 million names long.

that the day's heat didn't cause InSight's arm to droop while it was reaching out to place the heavy SEIS instrument.

The first measurements were drowning in so much noise that it would have been impossible to detect any marsquakes. Fortunately, the team anticipated this and on Sol 66 deployed the Wind and Thermal Shield over SEIS as protection from the elements. The top half of the shield was a dome, beneath which was a gold foil thermal shield that would help keep the temperature even. This was weighted down by a chain and scale mail skirt, which was heavy enough to keep the shield in place against the strongest of winds, but which could mould around pebbles and pits on the surface to create a good seal.*

The next instrument set up was the heat probe, HP3, which was placed on the surface on 12 February 2019. The team initially thought InSight had come down on a bowl of sand a metre deep, covering over Martian regolith for the next 8–12 metres before hitting solid bedrock. The probe should be able to drive itself into that kind of terrain relatively easily by winching up a weight, then dropping it suddenly to hammer the probe down. It was a laborious process, but it should only have taken a few months to hammer in the full 5m the probe was capable of, dragging its string of sensors behind it.

Of course, nothing is as straightforward in practice as it is in theory. The mole began hammering into the surface on Sol 92. In 4 hours, it hammered some 4,000 times. But something was wrong. The probe should have burrowed 70cm, but it seemed to have only gone in between 18–50cm.

When the team stopped hammering in early March 2019, HP3 had only burrowed around 30cm into the surface. In June, InSight removed the support structure of HP3 so that Ground Control could see what was going on. To their surprise, the probe had dug a hole twice as wide as they were expecting. The regolith wasn't giving the nail the friction it needed to hammer into the surface, meaning it was twisting around rather than burrowing. After InSight poked the hole with its mechanical arm, it seemed the soil had very little cohesion, and the probe was sliding around in it. To make matters worse, their early attempts to dig had

---

*   The scales were made by fantasy costumiers Ring Lord, who used similar scales to create the Elven armour in *The Hobbit* movies. Hopefully, SEIS won't have to protect itself from troll attacks as well as wind and heat changes, though.

compacted the soil below the nail's point, creating a hard crust it would need to punch through.

InSight used its robotic arm to push the heat probe into the side of the hole it had dug, hoping it would give enough friction to dig in deeper. They managed to get the main body of the probe completely buried this way, but after that point the arm could no longer reach the mole to offer its support. By the summer of 2020, this technique had managed to push the mole beneath the surface. But on 20 June the InSight engineers noticed that scoop was beginning to jiggle, suggesting that the probe was jumping up and hammering against the scoop rather than digging in any further. As of August 2020, they were still investigating the issue, and trying to work out what the next step should be.

The issue highlights one of the major drawbacks of robotic missions – if something goes wrong, you only have what's on the lander to fix it. If a human had been on the surface, they would have immediately seen the problem and either moved the probe somewhere it could get more purchase or helped poke it down in the current spot. What would have taken a geologist a few hours is taking InSight months.

Even without the heat probe, the rest of InSight is still hard at work. SEIS detected its first definitive marsquake on Sol 128. By Earth standards, it was a minor tremor, but it still qualified. As time went on, the instrument was detecting one or two quakes every day.

As I write this, InSight has only just begun its time on Mars. It will take many years for the probe to take all its data as it sits quietly taking in the Martian landscape, and it will probably take another few years after that until the full picture of Mars's interior becomes apparent. We'll just have to wait and see.

22

THE FUTURE OF MARS

Mars is unarguably the most explored of all the planets beyond Earth, a trend that shows no sign of stopping. In 2020, not one but three Mars rovers were due to head on their way towards the Red Planet, each with a different goal in mind.

After years on the drawing board, NASA is finally mounting a sample-return mission, starting with the Perseverance rover. Rather than being a scientific explorer in its own right, the rover will run around the surface, collecting samples and packaging them up in tubes. These will then be collected by a joint mission between NASA and ESA, which is due to fly in the 2030s.

Perseverance won't just be setting things up for a robotic mission, but a potential human one, too. The rover will carry the Mars Oxygen In-Situ Resource Utilization Experiment (MOXIE), which will extract carbon dioxide out of the atmosphere and split it apart to create oxygen. Although MOXIE will only produce 10g of oxygen every hour – about seventy breathes worth of oxygen* – the same technology could one day be scaled up to provide air for astronauts to breathe or fuel for rockets.

One more experiment that will probably earn its fair share of headlines is the Ingenuity helicopter scout, a 1.1m drone-like probe that will traverse the thin Martian air to explore a wide area. The instruments on Ingenuity

---

* Only around 5 per cent of oxygen is consumed in each breath, so I think 10g of oxygen could last for around 90 minutes of one person breathing normally, but I really would not advise trying it based solely on my (literal) back-of-the-envelope calculations.

are very simplistic, limited to the navigation systems and a camera, but if Ingenuity works, similar explorers could end up covering vast areas of the Red Planet.

Meanwhile, ESA and Roscosmos are collaborating on a mission to land a rover as part of the ExoMars project that began with the Trace Gas Orbiter (TGO) in 2016. While it is a highly accomplished probe in its own right, TGO will serve as a communications hub between the rover and Earth. The primary purpose of both the ExoMars Orbiter and rover is to investigate the possibility of life on Mars. To reflect this, the rover was named Rosalind Franklin – a chemist who was instrumental in the discovery of the double-helix structure of DNA – and will pick up NASA's tradition of all Mars rovers being female.

Rosalind Franklin's most unique asset is its 2m long drill. All previous explorations of Martian life have been stuck looking at the irradiated surface, but the ESA rover will be able to burrow down deep to a region where organic material has a much better chance of surviving. The rover will take samples from various depths to see how the composition changes the further down it goes.

Of course, this all depends on ESA making a safe landing. They had a dry run when TGO entered orbit in 2016, as it was carrying the Schiaparelli Lander. The lander was a mock-up of the system they planned to use with ExoMars and was developed alongside Roscosmos. However, when it attempted to land on the surface on 19 October 2016, Schiaparelli was lost during the final descent stages. The cause appeared to be a computer glitch, as the spacecraft began spinning rapidly after entering the atmosphere. The computer got confused about what altitude Schiaparelli was at and released the parachute early. Then, the thrusters only fired for 3 seconds instead of 30, as it thought it was near the surface when in reality it was still 3.7km in the air. The lander fell the rest of the distance, slamming into the ground at 540km/h. Ouch. Fortunately, as the problem was software related, it was a fairly easy fix.

This wasn't the only problem with the landing system. In May 2019, Earth-based testing highlighted another design fault, this time concerning the parachutes. The rover will use a series of four parachutes to slow down, but during a drop test, two suffered serious damage while being deployed. The team managed to identify the problem, and were on track to make the necessary adjustments when the COVID-19 pandemic hit the world in early 2020. With strict travel restrictions preventing engineers moving around Europe where they were needed, ESA conceded that the lander couldn't be fixed in time and pushed the mission back to a 2022 launch.

The Chinese National Space Agency (CNSA) are also sending their own mission, Tianwen-1, consisting of an orbiter and a lander with accompanying rover. It launched towards Mars on 23 July 2020, but China are taking a leaf out of the early Soviet space handbook and have released few details about the mission. From what they have said, it appears the spacecraft will be a Pathfinder like test, learning how to land and operate on the Red Planet.

As I'm writing this, both Perseverance and Tianwen-1 are speeding on their way towards Mars, and are due to arrive there in February 2021.

The path forward after this trio of rovers is a little more uncertain. NASA and ESA have agreed to a joint mission to pick up the caches being left behind by Perseverance. The plan involves multiple stages. First, a European-built fetch-rover will gather up the cached samples and return them to the lander. Here, they will be bundled up into a sample-return capsule, which will then launch into Martian orbit. They will then join up with a flight stage to return them to Earth. Once the samples have landed and been recovered, they will be transported to a secure facility to be cracked open and investigated.

Although ESA committed to the project in the triennial meeting of all the agency's member states in November 2019, such international projects have frequently been the victim of political manoeuvring. However, rovers have always made charismatic ambassadors for planetary exploration. With record-breaking drills, exciting helicopter drones and a way of producing oxygen into the mix, the trio of rovers heading to the Red Planet in 2020 and 2022 should help to keep Mars on the agenda long enough for the 'return' part of the sample-return mission to get its footing.

But, even after all this work, the samples returned will probably amount to less than a kilo of dust. While there is a lot that can be done with that, it's almost certainly not enough to answer one of the big questions that has run throughout this section – is there life on Mars? Even if there are, or were, microbes, Perseverance would have to be incredibly lucky to be in the exact right spot to pick one up.

It seems the only way humanity will get a definitive answer to the question of life on another world is to send a human to investigate. NASA's plans for a crewed mission to Mars have been running in the background of all its human and Mars missions for decades; for instance, one of the key reasons for developing the Skycrane was to learn how to deploy the heavy infrastructure needed for a human mission. Alas, the details of when and how these missions might happen are still excruciatingly vague.

Currently, the agency is focused on the International Space Station and building the Lunar Gateway – a permanent station around the Moon that will allow astronauts to travel to and from the surface. This will help the agency learn how to live and work away from the support of Earth, but there is still a long leap between operating at the Moon around 400,000km away to travelling to Mars 400 million km away.

Although the path forward isn't very clear at the moment, one thing seems to be obvious – humanity isn't done with the Red Planet.

PART 4
THE SOLAR
SYSTEM'S
OTHER SIBLINGS

23

TO VISIT A COMET

For the first few decades of the Space Age, by necessity planetary explorers focused on our nearest neighbours: the Moon, Venus and Mars. As technology progressed, space agencies began to turn their eyes to the further reaches of our solar system. With millions of asteroids in the belt, not to mention hundreds of moons around the outer planets, mission planners were spoiled for choice when picking the next targets that humanity would explore. However, there is one substance on which planetary geologists home in.

On Mars, the mantra has long been 'follow the water'. The philosophy is well deserved. Water is, arguably, one of the most interesting molecules in the cosmos. It's called the universal solvent, because more things dissolve in water than in any other known liquid. Once a substance has dissolved in water, its component parts are much more open to interacting with each other. It also allows them to travel with the liquid wherever it goes, until either the water evaporates, or the chemicals settle out.

Water's incredible dissolving power is why it's so important to life on Earth. It helps transport minerals around our bodies and enables the reactions that keep them running. However, it isn't just biological evolution that water has helped to shape but the planets as well.

On Venus, water building up in the atmosphere led to the runaway greenhouse effect that boiled it alive. Meanwhile, the absence of water on the planet's surface prevented the tectonic plates from sliding over each other. On Earth, tectonic activity is a major geological force. The process removes carbon dioxide from the atmosphere, regulating the temperature.*

On Mars, water shaped the surface, carving out canyons and smoothing out flood plains. Even today, water is frozen into the surface soil across the planet's poles, turning what could have been soft Martian regolith into hard permafrost. In the outer solar system there are entire moons made of ice, where the temperatures are so cold water begins to act like rock.

Nowhere is the effect of water more noticeable than on our own planet. Water is everywhere, with every form – solid, liquid and gas – working together to create the planet that surrounds us.

If we want to understand the terrestrial planets, understanding the history of their water is vital. But one question looms large over every investigation into water in the inner solar system. Where did it all come from? To answer that we have to go back to the beginning.

The solar system first formed from a pre-solar nebula, a cloud of gas and dust floating in space. Over time, the gas came together and eventually the cloud grew large enough that the Sun formed at its centre, while the rest created a swirling plane of gas called the protoplanetary disc, out of which the planets formed. For several million years, these planets drifted around the solar system, a process called migration, before finally settling into the balanced arrangement we know today.

Among the various chemicals within the pre-solar nebula was water. Here is where we come across gaps in our understanding. The Sun was much brighter in its youth, and the inner solar system was hotter. The frost line – the point past which surface water freezes – was much further out, around where Jupiter is now. When the inner planets formed, any water that condensed on their surface would quickly boil away and be lost to space.

The terrestrial planets might have formed further out, beyond the frost line, and then drifted inwards, but this seems unlikely. If the temperatures were cool enough for water to exist as a liquid, they'd also be too cool for

---

*   This is called the carbon–silicate cycle. Volcanoes belch out carbon dioxide into the atmosphere, which causes the planet's temperature to rise. Various knock-on effects mean carbon-rich rocks form faster, sucking carbon out of the air and making the global temperature fall. Tectonic plates carry these rocks back into the Earth's mantle, where they mingle with the magma, ready for a volcano to erupt, belching the carbon dioxide back into the atmosphere. This cycle takes hundreds of thousands of years. While it will eventually correct the carbon dioxide currently being released by humans burning fossil fuels, it will be long after the effects of climate change have ravaged Earth's biosphere.

the interactions needed to form a rocky planet. So they must have formed without any water on their surface.

One possible explanation lies in the fact that the planets weren't dry to their core. As the planetoids came together, they trapped water vapour within the rock, locking it away, deep down. The question then becomes how this water got to the surface.

There are two main theories, although the truth may be a combination of both. The first is that this water within the planet was brought to the surface, either through volcanic eruptions or meteor impacts cracking open the crust. The second is that water was brought from elsewhere in the solar system.

After the planets formed, there was a lot of rubble left behind. Some was in the form of rocky asteroids; some in the form of giant lumps of ice. In the outer solar system, it was cold enough for water to clump together, creating large bodies in much the same way that rocky ones formed in the inner solar system.

There are now millions of these objects encircling the outer solar system in a ring known as the Kuiper Belt. It's been theorised that there's an even larger cloud of icy objects beyond that, stretching away up to 50,000 times the distance between the Earth and Sun, known as the Oort Cloud. Occasionally, icy bodies from both these regions get thrown inward towards the Sun, becoming comets.

In these early days, there were so many asteroids and comets being hurled into the inner solar system that they regularly crashed into the growing planets. We can still see the scars of these collisions today in the craters left behind on the Moon. When these rocks crash-landed, their raw material mingled with the planet, including their water. Could asteroids or comets be responsible for bringing water to the planets' surfaces? While asteroids are much dryer than a comet made of ice, they do still have water, and there are many more asteroid impacts than cometary ones.

To work out whether these two delivery systems were responsible for bringing water to the Earth, astronomers needed to find out what 'flavour' of water asteroids and comets are. Water is made from an oxygen atom and two hydrogen atoms, H_2O. However, hydrogen has a heavier form called deuterium, which has an extra neutron in its nucleus. When water forms with one hydrogen atom and one deuterium atom, it's called heavy water. On Earth, around one in every 3,200 water molecules is a heavy water molecule, so a 500ml bottle of water contains around 0.2ml of heavy water.

The chemical difference between heavy and normal water is negligible, meaning the ratio between the two is the same as when it first arrived on Earth. This also means the ratio would be the same on whatever object brought water to our planet. If astronomers could find which object had the same ratio, then they would know what was responsible for Earth's beautiful oceans and rainy Thursday afternoons. And they started the search with comets.

Humanity has known about comets for millennia. When the celestial visitors pass close to Earth, their bright tails streak across the sky for months at a time. The earliest surviving mention is from Greek philosopher Aristotle, although the first description of a specific comet came from a Chinese astronomer in 240 BC, referring to one as a 'broom star' because of its bristle-like tail.

Their seemingly random appearances, sometimes generations apart, means much of what we know of these early comets is wrapped in myth. Shortly after the assassination of Julius Caesar in 44 BC, a comet was taken as a sign of the murdered emperor's ascension to godhood. By the Middle Ages of Europe, they were seen as 'the heavy hand of God' – bad omens heralding in death and disease.

The truth about comets only became apparent in 1705, when the English astronomer Edmond Halley deduced that a comet appeared in the sky every seventy-four to seventy-nine years. He was the first person to realise that these recorded sightings were all of the same object, which looped around the Sun every seven and a bit decades. He predicted the comet would return in 1758. Halley passed away in 1742, so never got to see his comet return. Instead, it fell to German astronomer Johann Georg Palitzsch to hunt it out, which he did on Christmas Day in 1758. Ever since, it has been known as Halley's Comet.*

Around 200 years later, in the 1950s, people found out exactly *what* it was that was in orbit around the Sun. By now, astronomers had noticed that comets have two tails, blowing out at slightly different angles, with one much brighter than the others. The discovery caught the attention of Fred Whipple, a former Iowan farm boy born in 1906, whose aptitude

---

*   It would later turn out that the 240 BC comet was an apparition of Halley's comet some 1,500 years earlier.

for mathematics led him to become a professional astronomer. Whipple had spent most of his career looking at photographic plates of the night sky, locating the distant glow of comets or the bright flash of a meteor captured forever in silver nitrate. He suggested that these long tails originated from a ball of water ice mixed with dust, known as a nucleus. As the comet neared the Sun, the ice melted, releasing two tails; one of dust, one of water. The press quickly gave this idea the much snappier title of a 'dirty snowball'.

Ground-based observations and theorising could only tell astronomers so much. If they really wanted to know what a comet was made of, they were going to have to get up close. Luckily, the wait wasn't long: Halley's Comet was due back in the mid-1980s, its first trip into the inner solar system since the dawn of the Space Age. This would be a once in a lifetime event and every space agency wanted to take advantage of the opportunity to get a close look at a comet. In the end, eight different spacecraft examined the comet, a fleet known as Halley's Armada.

The spacecraft were built by several different nations working closely together. After one team made their observations, they passed their data onto the next team to help them get closer to the nucleus and build a full picture of the comet.

Three spacecraft took a look at Halley's comet from a distance. NASA's geriatric Pioneer 7 – launched in 1966 and now in solar orbit – was pulled out of retirement to image the comet as it passed the Sun. Then the Pioneer Venus Orbiter, also from NASA, imaged the comet as it passed by.

The International Cometary Explorer, a joint effort between NASA and the European Space Agency, which originally investigated the relationship between the Sun and Earth, made more of an effort to approach the comet. It had previously swung by comet Giacobini–Zinner, passing just 7,800km from the nucleus. It later flew through Halley's tail, but it never got closer than 10 million km from the comet's nucleus.

Five spacecraft got in closer. The Japanese space agency sent two probes: Sakigake, which tested out the technology needed, then Suisei, which used those findings to approach to 151,000km from the nucleus. Both Vega 1 and 2 flew a just a few thousand kilometres away from the comet after dropping their balloon probes off at Venus.

Finally, there was the European Space Agency. The fledgling organisation had been set up less than a decade before and had yet to undertake
a deep-space mission beyond the reaches of low-Earth orbit. Halley's
Armada gave Europe a chance to plant its flag and take centre stage at
a huge space event. Its standard bearer would be the spacecraft Giotto,
named after the Italian Renaissance painter who used the comet as inspiration for the Star of Bethlehem in his *Adoration of the Magi*.

The spacecraft's main purpose was to get in close and take colour footage of the coma (the cloud of gas and dust around the comet's nucleus)
as it passed a few hundred kilometres away. Meanwhile, other instruments
would analyse the composition and flow of the gases from the comet.
Critically, the spacecraft would be able to measure the isotopic ratio of the
comet's water.

Although many people wanted spacecraft to land on the comet, Halley
would be moving incredibly fast as it sped past the Sun. Measuring just
11km across, it had barely any gravity to pull objects towards its surface.
Meanwhile, the gas streaming out would push against anything that came
close. With the Space Age just thirty years old at this point, a landing mission was far beyond any agency's ability.

But while Giotto couldn't reach out and touch the comet, the comet
could reach out and touch the spacecraft, with potentially disastrous
results. Along with gas, Giotto would be pummelled with dust shed by
Halley – dust that had destroyed 80 per cent of the Vega spacecrafts' solar
panels when they flew past. The spacecraft and the comet (and the dust
coming off it) would be heading towards each other at 245,000km/h. At
that speed, a dust particle weighing just 0.1g, around the size of three
grains of rice, would tear through 8cm of solid aluminium.

An 8cm shield of aluminium would have added 600kg to the spacecraft's weight, making it far too heavy to launch. Instead, the team used
a two-layered shield invented by Whipple. First, the particles would hit a
1mm aluminium sheet. The smaller particles would vaporise against this
layer, while larger debris would punch through, losing some energy in the
process, and reach a second layer of 12mm-thick Kevlar. The two layers
combined would protect the spacecraft from all but the largest pieces
of debris.

On 13 March 1986, Giotto was ready to fly past Halley. Vega 2 had
flown by five days beforehand, allowing Giotto's flight planners to pinpoint the comet's position to within 40km. The ESA team could now
refine the spacecraft's flight path to approach to 550km from the nucleus –

as close as they dared, given the dust seen by the Vegas. When Giotto was 122 minutes away from the expected point of closest approach, the spacecraft felt its first impact. Dust hit the spacecraft for the next 2 hours, the rate sharply rising as it passed through a jet of gas from the comet.

Back at the European Space Operations Centre in Darmstadt, Germany, ten different teams scrutinised the spacecraft's data as it streamed in. The images on the view screens revealed that the comet was shaped more like an elongated potato than the round ball they'd expected. They eagerly watched the clock, counting down the minutes and seconds until closest approach when, 7.6 seconds before the big moment, the images on the screens wobbled, then went blank.

Those watching feared the worst when snatches of information began to come through. Whatever was wrong with Giotto, it was alive. It didn't take long to work out what had gone wrong. Moments before closest approach, Giotto was struck by a fragment of dust that weighed 1g and was the size of a pea. The Whipple shield had done its job keeping Giotto alive, but the impact knocked the spacecraft sideways, directing its antenna away from Earth while puncturing the main body and damaging vital systems. Giotto's recovery systems kicked in, restoring full contact half an hour later, but by this point it had already sailed past the nucleus.

The spacecraft registered its last dust impact 49 minutes after closest approach. It had endured 12,000 impacts and come within 596km of the comet – further than intended, but still ten times closer than any spacecraft had come before. Humanity had glimpsed the heart of one of the most mysterious objects within our solar system.

Despite the last-minute hiccup, the Giotto mission was deemed a success. The images showed the peanut-shaped body of the comet's icy nucleus, haloed with gas and throwing off roughly 3 tonnes of material per second. But rather than coming off in a uniform cloud, the gas was being thrown out in powerful jets. The chemical measurements showed that the rock was mostly made of water ice, just as expected, with traces of carbon compounds mixed in. Their dirty snowball was as black as coal.

The amounts of trace chemicals in the rock lined up with theories about the composition of the cloud of gas that had formed the Sun, indicating that the comet originated from the earliest days of the solar system. Importantly, the isotopic ratio of the water was the same as that geologists think was present in the pre-solar nebula. However, this meant Halley contains twice as much deuterium as the water on Earth does. Water definitely wasn't brought to Earth by a comet like Halley. Now astronomers needed to find out if Halley was typical. Or did other comets have a different mix?

One fact Halley's Armada resolutely proved was that a comet was never going to be an easy target to approach. Unlike planets, comets didn't sit idly by while humanity's robotic emissaries invaded. Comets fought back.

As usually happens when humanity first visits a new place, the Armada left us with more questions than answers. Although the fleet proved first-hand that comets were made of water, they also highlighted how many secrets were hiding within these icy visitors. It was a tantalising first taste of what could be learned from these transient worlds, but if space agencies wanted to get their teeth into a comet, they were going to have to get closer.

24

CHASING A COMET'S TAIL

When the Armada flew, the United States had only been able to watch from a distance. Despite having an up-close comet mission on the books since the late 1970s, a combination of politics, bad luck, budgeting and scientific infighting meant that they couldn't quite pull it together in time for Halley. For the first time, NASA had missed out on a major space endeavour. To make matters worse, the Armada missions came a few months after the Challenger disaster.

In January 1986, the world watched as the Space Shuttle broke up during launch, killing all seven astronauts, including schoolteacher Christa McAuliffe who had joined the crew as part of the highly publicised Teacher in Space Project. The American people were beginning to question whether spaceflight was worth the cost – not just in dollars, but in human lives.

While they might have missed making a trip to the headline comet Halley, NASA hadn't given up on comet exploration. If they couldn't be among the first, then they would just have to be the best. The agency began to put together the Comet Rendezvous Asteroid Flyby (CRAF) mission, an impressive flagship spacecraft that would not just fly past a comet, but follow alongside one, coming in as close as 6km. Most excitingly, it would carry a penetrating lander – basically a 1.5m long golf tee that would burrow into the surface – allowing NASA to investigate the comet's physical and chemical properties from the surface itself.

By 1989, the project was combined with the Cassini mission – an ambitious spacecraft being sent to Saturn that would also drop a probe onto Titan (see Chapter 29). The two spacecraft would be based on a new base design for NASA's next generation of planetary missions, the Mariner Mark II. The joint project was huge, with both halves shaping up to each be billion-dollar missions, but by linking them NASA hoped to save $500 million.

Unfortunately, the project rapidly went off the rails. In an effort to reduce costs, both missions were scaled back. In 1990, the penetrator was removed from the mission after its estimated cost quadrupled. Without it, the scientific impact of CRAF was seriously reduced.

In 1992, Goldin's 'faster, better, cheaper' era began. Both CRAF and Cassini were just the kind of bloated, expensive mission to be avoided under the new philosophy. As NASA's budget was squeezed, CRAF-Cassini was one of the first projects up for review. Fearing that continuing both CRAF and Cassini would result in the cancellation – or worse, failure – of both, Congress gave CRAF the chop in the 1993 budget.

The scientific community decried the decision. The mission could have been scaled back to keep within budget while remaining scientifically valuable, but the consensus amongst the science teams was that the mission had been cut with no real effort to discover if this was possible. NASA was letting itself be led by budget and politics, rather than what was scientifically best.

Never ones to let their hard work go to waste, several of the CRAF teams took the ideas of the large mission and worked on plans for low-cost missions that would mesh better with the new ethos at NASA. In August 1994, the floor opened for proposals for the next round of hundred-million-dollar Discovery-class missions.

Cometary missions were a hot topic of the day, with seven out of twenty-eight proposed missions intended to head towards these celestial visitors. After fighting off the competition, two missions were selected, one of which was Stardust – a plan to capture a comet's tail and return it to Earth.

The mission, proposed by Donald Brownlee from the University of Washington, was simple, cheap and had a high probability of success – a very unusual set of descriptors for a sample-return mission. It also didn't hurt that the romantic idea of 'capturing a comet's tail' would go over well with the members of Congress who held NASA's purse strings.

The concept was deceptively simple. The spacecraft would sweep through a comet's tail, collecting dust particles as it went. These would then be sealed away and returned to Earth. Once here, the particles could be analysed to find out exactly what, other than water, makes up the dirty snowballs of our solar system. Behind this basic idea, however, was a complicated engineering problem. How do you capture fragile particles of dust and ice travelling at thousands of kilometres per hour without destroying them? It was like trying to catch a bullet made of glass without breaking it.

The answer was a strange, largely unknown substance named aerogel. A wet gel is basically a 3D structure that traps water into a semi-solid form. By removing the water without destroying this structure, chemists created aerogel. Made of 99.8 per cent air, it is a strange thing to behold, like blue smoke trapped into a solid shape. Besides its unnatural beauty, aerogel is an excellent insulator, can survive high temperatures and is exceptionally strong. When it was first discovered in 1931, it was also expensive and difficult to work with, so it remained little more than a curious invention for the next fifty years.

For the Stardust team, however, it was perfect. Aerogel had just the right balance of density; soft enough for particles to burrow in without vaporising on contact, but dense enough to stop them passing through entirely.

The aerogel was held in a collector that would waft through the comet's tail as Stardust flew through it. Officially, this was known as the Sample Return Collector, more commonly nicknamed the 'tennis racket', due to its size and shape – a roughly circular panel separated into 130 2cm by 4cm compartments that housed the aerogel. Each of these compartments contained a 3cm-deep block of aerogel on one side, which the cometary particles would get embedded in, and a 1cm block on the other side to capture particles from beyond our solar system.* The density of the aerogel increased with depth. The 'fluffier' particles would stick in the low-density aerogel at the front without being destroyed on impact. Meanwhile, the more solid particles would shave off most of their velocity on this outer layer, allowing the denser material deeper down to stop them. Early tests found that particles travelling as fast as 7km/s could survive impact.

Stardust's construction by Lockheed Martin went smoothly, with none of the drama that would plague many other planetary missions. The design

---

*   Stardust collected these during its five-year journey to the comet. The trails these left behind were microscopic and required going through 71 million microscope images to find. The team passed the task to members of the public via the Stardust@ Home project, where thousands of volunteers searched for signs of a potential particle. The first candidates were found in 2010 by Canadian Bruce Hudson and Brit Naomi Wordsworth, who got to name their particles – Orion and Hylabrook. In August 2014, the team announced they'd removed seven particles, and that they were indeed from another star.

of the main spacecraft was fairly standard – a box-shaped main body flanked by two long solar panels. A cap on the spacecraft's nose covered over the sample collector for most of the journey, only opening to let the tennis racket pop up when Stardust passed the comet, before closing over the collector afterwards to seal away its precious samples. It seemed that 'faster, better, cheaper' was paying off. The spacecraft sailed through testing and on 7 February 1999 it launched from the coast of Florida on time and on budget.

It took another five years before Stardust was ready to do what it was built for. Its destination was comet Wild-2 (pronounced Vilt). The comet was considered young – not because of its age (some 4.5 billion years) but because it was new to the inner solar system. In 1974, Wild-2 had flown close to Jupiter. The gravity of the planet shifted the comet's orbit, throwing it towards the Sun. While comets like Halley have been warmed by the Sun hundreds if not thousands of times, Wild-2 had only passed near to the Sun five times. If comets were the pristine material left over from the planets' formation, then Wild-2 was as pristine an example we're ever likely to find.

In 2002, the spacecraft broke its journey with a quick flyby of asteroid 5535 Annefrank, primarily as an engineering test of its cameras, which were more of a navigation tool than a science instrument. The spacecraft's memory could only store seventy-two pictures at once, which was about as many as it would have time to take during the brief comet flyby, and the team wanted to make sure they made the most of them. The main observations of the comet would be done from Earth, where the world's best telescopes would be examining every aspect of the comet they could.

On Christmas Eve 2003, Stardust was finally closing in on the comet and deployed its detector, ready to waft it through the tail. Knowing what had happened to Giotto, the team decided the spacecraft should hang back 237km from the comet when it passed by on 2 January 2004. Even so, as it neared the comet, Stardust bravely weathered the bombardment of particles as new jets burst from the comet, pelting the spacecraft with a cloud of dust. It made it through – battered but alive – and 5 hours later the collector was stowed, ready to return to Earth.

Two years later, on 15 January 2006, Stardust came back down to Earth. The atmosphere in the control room as they monitored the descent was particularly tense. Just a year earlier, Genesis, a similar mission to collect solar wind particles, failed to deploy its parachutes during re-entry. It slammed into northern Oregon at 311km/h, rupturing the capsule, contaminating the samples and rendering them useless.

Stardust was spared the same fate. The parachutes deployed and it came down, intact, on the US Air Force Test and Training Range in Utah. As comets were deemed unlikely to host any potentially hostile microbes, the samples didn't require any stringent planetary-protection procedures. The container was transferred to the Johnson Space Center, Texas, to find out how the aerogel had fared.

The blocks had survived their trip and were now pitted with thousands of carrot-shaped burrows from dust penetrating the aerogel. As the researchers took a closer look, they were stunned to find that at the end of some of the tunnels was a rocky particle, suggesting that Wild-2 was slightly rocky, more like an asteroid than a classic comet. It seemed the story of Wild-2 was more complicated than they'd expected.

A slice of Stardust's aerogel showing the carrot-shaped burrow left behind by a particle from a comet's tail. (NASA/JPL-Caltech/University of California, Berkley: photojournal.jpl.nasa.gov/catalog/PIA02189)

While these particles were easy to remove, examining the more fragile material would be a far greater challenge. Despite their efforts, the particles had broken apart in the aerogel and the volatile elements (those that boil at room temperature) had long since melted away. However, there was still enough to begin analysing what makes up a comet, and the team extracted what fragments there were making a note of the shape the burrows had made to build up a picture of what the fragment had looked like before striking Stardust. In the end, they found over 10,000 particles buried in the aerogel, and even managed to recover some of the residue left behind by the volatiles that had long since boiled away.

The results indicated that the comets were very different from meteorites – the only samples of space rocks available at the time – but they did resemble the dust found in the space between planets.* Analysing what water there was revealed that, just like Halley, Wild-2's water was very different to that of the inner solar system, with deuterium levels three times greater than those found on our planet.

It was seeming like comets might not have brought water to Earth – but they might have brought another ingredient of life. Amongst the cocktail of chemicals found in the grains was some organic matter, most notably the simple amino acid, glycine. There were initial worries this might be an Earth contaminant, but the amount of carbon-13 suggested it hailed from another world. Could a different comet have brought the basic building blocks of life to Earth?

The particles seemed to be a mixture of soft and rocky grains, glued together by this organic material. This was unusual. The rock must have formed at high temperatures, yet the icier parts formed at low ones. How could a comet be both hot and cold? The answer still evades explanation.

The rocky material may have formed in the inner solar system, but then been swept out later; or there could have been localised hotspots in the outer solar system that made rocky material; or Wild-2 might have just been a strange aberration. One thing was certain, though – the line between comet and rocky asteroid was far more complex than anyone had anticipated.

---

*   Interplanetary particles can be harvested from high in Earth's atmosphere by planes, making them a much cheaper to study than mounting a mission to a comet.

Stardust had been sent out to capture a comet's tail and now planetary scientists had a much better idea of what goes into making a comet. With the sample container dropped off at Earth, that job was done, but the main body of the spacecraft continued on. It had enough fuel to rendezvous with another target and in July 2007 NASA announced it would fly past another comet, Tempel-1. Stardust-NExT (New Exploration of Tempel-1) would be limited in scope as the instruments were designed to support a sample collector the spacecraft no longer had. But half a mission was better than none.

The dust analyser was still working, while the particle flux instrument could measure how many particles struck the spacecraft. However, the measurements that would really come to the fore were the seventy-two images the camera could take. Stardust might have been the United States' first cometary mission, but it wasn't the only one it had mounted, nor would this be the first time humanity had visited Tempel-1.

25

DEEP IMPACT

In 1998, NASA's fourth round of Discovery missions were up for selection. Among them was an intriguing new concept that would attempt to make contact with a comet, on purpose this time. Improved Earth-based telescopes meant astronomers knew the paths of comets much more accurately, so flight planners could plot a trajectory to get in close to a comet. Then, advancements in automated navigation would allow a lander to guide itself in to the comet.

Placing a soft-lander on the surface, however, would first require matching the comet's speed. That was far beyond the scope of what a Discovery class mission was capable of. However, it could fire a hard-lander into the surface. When this impacted, it would carve out a crater that would not only punch through to the pristine ice beneath, but would also kick up a cloud of debris that the flight stage flying past could get a look at.

The name of the proposed mission was, appropriately, Deep Impact.* By looking at the crater and cloud, the mission would be able to garner information not just about the composition of the comet, but its internal structure too. Just like Stardust, it was low cost, had the potential for some revolutionary cometary science, and had the bonus of being an eye-catching mission for the public. NASA were going to try and shoot a comet! In 1999, it was selected to be the eighth Discovery mission.

There was one big drawback – it was risky. If NASA underestimated the size of the impactor, the crater might be too small to see; too large and it

---

*   Entirely coincidentally, a film of the same name was released in 1998, about an asteroid crashing into Earth.

could shatter the comet entirely. And, of course, there was a danger they might miss the damn comet in the first place.

The goal was to approach the comet as slowly as possible, giving the spacecraft time to aim before the impactor and flight stage split apart. The impactor would be a fully functional, battery-powered spacecraft that could operate on its own for 24 hours. It would need to be as smart as the cruise stage to run the navigation software. Although the cruise stage would be able to release the lander in the rough direction of the comet, the lander would need to use sophisticated navigation software, cameras and thrusters to guide itself in to the comet. This would also be able to compensate for any jet of gas that might knock the impactor off-target at the last moment.

While most spacecraft frantically try to keep their weight down, Deep Impact would actively increase its own, as the more mass it had, the more energy it would strike the comet with, creating a bigger crater. Half of the impactor's 372kg mass came from a huge lump of copper, chosen because NASA didn't expect a comet to contain any, meaning they could easily remove its signature from the results. The copper was machined into a disc about a metre wide, which was fixed onto the bottom of a circular instrument deck. The instruments where bolted to the other side of the deck, clustered in the centre to keep the balance, before being covered over by a protective shield. The end result was a blocky lander that looked every bit as bulky as it was built to be.

The flight stage would monitor the whole event from afar. Rather than just being the mule to carry the impactor, the flight stage was an advanced monitoring station with two cameras. First, a high-resolution camera would image the comet in the lead-up to the mission to map out the comet's shape, rotation and composition to pinpoint the perfect place to aim the impactor. Then, during the impact itself, these cameras could observe the cloud of ejecta as it bloomed and fell.

Then there would be a medium-resolution camera. During the run-up it would use its infrared spectrometer to study the composition of the comet, but its real purpose would come to the fore during the impact itself. The encounter would only last around 800 seconds; not much time to take images. A lower resolution meant a faster read rate, so the camera could capture images much quicker than its higher-resolution cousin. The camera wouldn't be able to pick out the fine detail, but it would give a better overview of how the cloud changed over time.

In both cases, it was the time evolution of the cloud that would yield some of the most interesting data. How the dust cloud grew would reveal the gravitational pull of the comet, helping to understand its internal structure. How much gas escaped would uncover the amount of ice locked away within the comet. The first gas and dust released would be from the material at the surface. As the impactor penetrated deeper into the comet, it would kick up debris from ever further down. By watching the way the cloud evolved over time, the team would really be seeing the different layers of the comet.

Once the dust cleared enough, Deep Impact could image the crater left behind. The size of the crater would depend on the comet's physical properties. If the comet was soft or loosely packed, then the penetrator would plunge deeper beneath the surface. With most of its energy spent driving in deep, the crater would be smaller. Conversely, if the comet was hard or tightly packed together, then the impactor would be stopped much more rapidly, and that same energy would instead create a much bigger crater.

Finally, 800 seconds after colliding with the comet, Deep Impact would be forced to turn away. The blast would create a hail of debris heading straight for the spacecraft travelling at 10.2km/s, forcing it to hide behind its Whipple shield, while continuing to transmit its data back to Earth in safety.

On 12 January 2005, the spacecraft launched from Cape Canaveral. Now came time for the six-month long cruise towards its target comet. The team had been forced to consider several criteria when deciding where to aim Deep Impact. Firstly, the comet had to be big enough – at least 2km in size. Any smaller and they risked missing it. Next, it had to be visible from Earth so terrestrial telescopes could watch along as the impact occurred, just as they'd done during the Stardust's flyby. Finally, the spacecraft needed to be able to approach from an angle and speed that would allow the flight stage enough time to see the impact and transmit data home.

The comet that fitted all these needs was Tempel-1. If everything went to plan, then the impactor would hit home on 4 July 2005. What could be better than cosmic fireworks on Independence Day?

Even though the comet had been discovered in 1867, Tempel-1 hadn't been studied much since. In the run-up to the mission, telescopes the world over were pressganged into observing the comet's size, shape, rotation, how reflective it was and its potential composition, all of which the Deep Impact flight team needed to know if they wanted the mission to be a success. They found the comet was approximately 6km long and rotated once every 40 hours or so. The observations gave a baseline from before the impact to compare to. The spacecraft could take the best, crispest images ever seen during the encounter, but they would be meaningless if the team didn't know what the comet had looked like beforehand.

Professional telescopes weren't the only ones looking at Tempel-1, though. NASA invited amateur astronomers – many of whom have equipment and skills that would make universities jealous* – to image the comet as part of the mission's extensive outreach programme.

The spacecraft itself first spotted the comet sixty-nine days out from impact, at a distance of 64 million km – the icy boulder it had come so far to see. Nine days later, on 5 May 2005, Deep Impact entered its approach phase, dedicating all its attention on the comet. Combined with the extensive ground observations, the team made sure they knew everything there was to know about Tempel-1 before it came time for the encounter. As it got closer, and the position of the comet was pinned down, Deep Impact performed several small trajectory corrections so that it was dead on target.

In late June, Deep Impact saw huge outbursts from the comet, as the Sun's heat caused new jets to spring up and throw ice and dust into space. From Earth, these jets are so short-lived they'd never been caught in the act of erupting, but with so many telescopes trained on the comet, observers finally got the footage of the jets they'd always wanted, and discovered they seemed to be getting bigger the closer the comet got to the Sun. The addition of steering thrusters to the impactor was seeming an increasingly wise decision.

Five days from the encounter, the spacecraft switched into impact mode. The team ran through every detail they could think of – from ensuring the camera exposure was set correctly to checking the spacecraft was on target. They would only get one shot, after all.

On 3 July at 6:07 a.m. GMT, the team released the impactor. Twelve minutes later, the flight stage burned to get out of the comet's way, slowing down to a much more camera-friendly 0.1km/s, relative to the comet.

---

*   I often wish there was a better word for home astronomers than 'amateur', because some of these people (either on their own or by clubbing together) manage to create world-class observatories using their own money (which can easily run into six figures). They set everything up themselves, often making their own equipment along the way. It's not uncommon for these astronomers to use their incredible facilities to assist professional scientists, and they do so for no other reason than wanting to further humanity's knowledge of the universe. It's astounding.

Meanwhile, the impactor was racing towards the comet at 10km/s, performing three course corrections to stay bang on target. The mission had a 99.9 per cent chance of hitting on target, but as the impactor drew ever closer to its target, the flight controllers couldn't help but worry about that pesky 0.1 per cent. After flying 430 million km, would Deep Impact miss its mark?

As the distance closed between them, the impactor's camera picked up more and more detail, with craters and crags appearing as the comet rushed up to meet it. Then, at 5:52 a.m. GMT on Independence Day 2005, 24 hours after leaving the cruise stage, Deep Impact finally had its deep impact.

The flight stage watched as the cloud bloomed out of the surface, but it would be another 5 minutes before Deep Impact sent the first images back to Earth, letting mission controllers know they'd hit their target. The spacecraft continued taking data for 14 minutes after the impact, before the risk of flying debris became too much and it was forced to adopt a defensive stance. Now safe, Deep Impact began the long task of transmitting all of the 4,500 images it had taken.

The images showed there had been a bright flash of light when the impactor had struck the comet with the energy of 4.8 tonnes of TNT. They had hit exactly on target – the impact had been successful! Perhaps too successful – the cloud of debris was so large it blocked out the view of the crater.

Deep Impact's flight stage was already on course to visit another comet, Boethin, so it wouldn't be able to return to Tempel-1 to see the crater. Fortunately, it wasn't the only comet-observing spacecraft in orbit around the Sun. Stardust-NExT just so happened to be looking for a new target and had enough fuel to head towards Tempel-1.

In keeping with Stardust's romantic nature, the spacecraft flew past the comet's heart on Valentine's Day 2011 (in United States time), taking its seventy-two images of the comet's newest crater. Almost six years after the initial impact, the team were finally able to get their first clear image of the crater.

After encountering Boethin, the flight stage met up with a third comet when it came within 700km of Comet Hartley 2 in November 2010. The spacecraft was due to have another encounter in 2020, but Ground Control lost contact with the probe in August 2013, bringing the mission to an end.

Comet Tempel-1, 67 seconds after being struck by the Deep Impact. (NASA/
JPL-Caltech/UMD: solarsystem.nasa.gov/asteroids-comets-and-meteors/
comets/9p-tempel-1/in-depth)

With the combined data of not just one, but two flybys, the Deep Impact
team were finally able to build a real picture of what a comet – or at least
Tempel-1 – was like. It was much dustier than they'd expected, which is
why there had been such a huge cloud of dust. The surface was surprisingly dry, with just 1 per cent covered in water ice. However, there didn't
seem to be a lack of ice beneath the surface, as the impact threw off an
estimated 250,000 tonnes of water (about 100 Olympic-sized swimming
pools). It seemed the gas of the coma and tail was coming from ice below
the crust, but probably not far beneath. This would help to explain why
comets are dormant until they come near to the Sun – it's only then that
the Sun's radiation can penetrate deep enough to reach the ice.

From the size of the flash, researchers were able to work out that the comet was around 75 per cent empty space. Rather than being a solid lump or rock or ice, it appeared to be more like a wet, icy sponge.

The close-up images from the impactor revealed strange cliffs and over-hangs that would collapse under their own weight on Earth. The low gravitational pull on the comet meant they could suspend themselves, creating a landscape that looked impossible to the terrestrial eye. Stardust's later observations revealed the impact crater was around 150m in diameter, about what they had expected given the amount of material thrown out.

The comet appeared to be a similar mix of the organic material seen at other comets. One of these chemicals, ethane, was much more prevalent around the comet after the impact. Either the interior contained more ethane than the surface, or Deep Impact happened to strike a pocket of ethane. Unfortunately, the only way to check which was true would be to go back and throw another impactor at some other part of the comet – a prospect unlikely to happen soon.

One surprising find was that the comet contained clays and carbonates, both of which require liquid water to form. All the water at comets we've ever seen has gone straight from solid ice to gaseous vapour. Exactly how these came to be on the comet – whether Tempel-1 was once partly molten or they'd formed somewhere, or somehow, else – is still unknown.

But perhaps the biggest benefit of the mission was its publicity. Deep Impact was a hit with the world's news, appearing on mainstream media across the world,* who were captivated by the story of how NASA had reached, and touched, not just a comet's tail, but its heart.

---

*   Not everyone was enamoured with the mission. Russian astrologer Marina Bai attempted to sue NASA for \$310 million, stating that the mission had disrupted the balance of the solar system, altering her horoscope by changing the course of Tempel-1. The comet was important to her as her grandfather had wooed her grandmother by showing her the comet. As the impact had changed the comet's trajectory by 10m – far smaller than the error margins of most known orbits – her case was thrown out.

26

ROSETTA

While Deep Impact had unlocked some of the secrets hidden beneath a comet's surface, these only hinted at the greater knowledge that could be gained by a closer look. What scientists needed was a longer, more in-depth view of a comet flying alongside and perhaps even the opportunity to set a soft-lander down on its surface. Such a mission would give a complete overview of what a comet was, outside and in. Fortunately, such a mission was already being planned by the European Space Agency (ESA).

Prior to 1984, ESA had only launched three space missions. Although their first major science mission, Giotto, would soon be on its way to Halley, they wanted flagship missions to cement their place as a major space power. They set out the Horizon 2000 programme, a long-term strategy that would set the agency's scientific programme for the next sixteen years.

The plan allowed for a handful of ambitious projects with budgets to match. To reflect their importance to the agency's scientific exploration, these became known as Cornerstone missions. The projects covered the gamut of astronomical research – the first two successful missions were the Solar and Heliospheric Observatory, SOHO (which is still monitoring our Sun twenty-five years later) and an X-ray observatory XMM-Newton (which looked to the very edges of our universe).

In 1992, teams from all over Europe began to submit their suggestions for the third Cornerstone mission. Amongst them was a group that had looked over to the CRAF mission that NASA had abandoned not long before. By scaling back CRAF, they created a mission that would rendez-vous with the comet while it was still in the cold of deep space, watching as it warmed up during its approach to the Sun, through its closest pass and then back out.

The spacecraft would aim to get within just a few kilometres of the comet's surface before dropping a lander. This surface probe would examine the comet as it changed during its journey around the Sun to create as full a picture of these wandering lumps of ice as it is possible to get, helping to pull apart the story of how water in the early solar system came to form comets. The mission was called Rosetta: as the Rosetta Stone had been the key to unlocking hieroglyphics, Rosetta the spacecraft would unlock the secrets of comets and the origins of our planetary neighbourhood.

The mission was just the right amount of daring and science for a Cornerstone mission. It was selected by ESA in 1993. Europe was heading back out to a comet. With an initial launch date of 2002–04, the team hunted for a comet to visit and initially settled on 46P/Wirtanen, a 1.4 km-wide comet that circled the Sun once every five and a half years.

A mission of this scale required a heavy spacecraft that would need a powerful rocket to launch. ESA only had one rocket up to the task – the Ariane 5, which was still being developed, but was due to be ready long before the 2003 launch date. Then, during the first flight of an Ariane 5 in 1996, the rocket exploded, destroying the $370 million spacecraft on board.

The cost of Rosetta was expected to run into the billions. ESA couldn't risk putting something that expensive on top of a dodgy rocket. The agency pushed the Ariane 5's ready date back by two years while they investigated the flawed rocket. The delays filtered down through the launch schedule, and it became apparent that even if Rosetta was ready by 2003, the rocket wouldn't be. They were going to miss the launch window for 46P/Wirtanen.

Fortunately, there were several other comets that would fit the bill. One of them, 67P/Churyumov–Gerasimenko,* would be in the right place for a launch date in February 2004. Just like Deep Impact's Tempel-1, it was a young comet. It had been discovered in 1969, although it appears to have had an encounter with Jupiter in 1959 that threw it into the centre of the

---

*   Pronounced Chur-eh-u-mov Geh-rah-sim-en-ko. I, like most people, will refer to it as 67P, because its full name is just as annoying to read over and over again as it is to type.

solar system. It looped around the Sun once every 6.45 years, getting as close as 1.3AU (where 1AU is the distance between the Earth and Sun).

The drawback to 67P was that it was bigger than Wirtanen, measuring around 4km long. It would be trickier to land on as the larger gravitational pull would accelerate the lander during the approach, increasing the chance of it crashing or bouncing off, but the Rosetta team were confident they could manage it.

Now they had their comet and the rocket, it was time to build Rosetta. Before any hardware could be constructed, however, some careful politicking was called for to decide who would get to build what, as the 'European' part of ESA's name made itself felt.

Large space missions these days are almost always international affairs. On most NASA missions there will be at least one instrument built – and more importantly paid for – by another nation. The larger nation gets the advantage of discounted hardware and the smaller ones can be a part of the next big scientific endeavour without having to foot the lion's share of the bill.

International co-operation is hardcoded into the ESA's DNA. The organisation is a consortium of over twenty different member states, pooling their funding to create bigger missions than they could ever mount alone. With that funding comes the understanding that each nation will be involved with building those missions, receiving the cash and cachet that comes with it.

If you look at any ESA spacecraft, you will find that every instrument is built by a different country. Look at each instrument and you'll probably find that the component parts started life in different parts of Europe as well. Each nation gets to build a proportion of the spacecraft in line with how much money they contributed to its creation.

This isn't much different from how NASA operates, as it contracts out most of the construction work in an effort to support the United States' space industry. It's just that while NASA only supports American businesses, ESA's obligations cover all of Europe.

Teams from the UK, Germany, Italy, France, Austria, Switzerland, Sweden and even the United States all oversaw the creation of instruments on board Rosetta. More countries were involved in putting each together, and more still would help to pick apart Rosetta's findings once it got to 67P. More than fifty industrial contractors from fifteen nations were involved in construction. Safe to say, it was a logistical nightmare.

Together, these instruments would analyse every aspect of the comet chemical and physical structure. Cameras would image the comet at a range of wavelengths, while spectrometers would pick apart what the comet and its tail were made of. Dust instruments would measure how much material was coming off the comet, and what it looked like. Radio experiments would help to pierce the comet's icy nucleus and find out what it looked like under the surface. By interpreting these, planetary scientists would be able to tell how the comet had come together, gaining a window into the early days of our solar system. Powering all this would be two enormous solar panel arrays, giving the spacecraft a 32m-long 'wingspan'.

The most ambitious instrument was the lander, named Philae after the lesser-known obelisks that helped linguists to decipher the Rosetta Stone. The mission planners wanted to put Philae down when the comet was 3AU away from the Sun, so the lander could look at the comet before it started becoming active. Philae would be relatively small, around the size of a washing machine. Thankfully, electronics had been continually shrinking for decades, meaning that the lander could pack far more into its diminutive form than would have been possible even ten years before.

Philae was designed with cameras to look at its surroundings, including a microscope and an infrared camera. It would carry an APXS to look at what the surface was made of, while an on-board chemical lab would analyse the chemistry of ice collected by a drill. All of this was contained within a hexagonal box, flanked on all sides by shining blue solar panels. This would be perched on three legs that looked far too slender to support its 98kg weight against the gravity of Earth, but which would be more than stable enough in the near non-existent pull of a comet.

While the orbiter aimed to gain a full overview of the comet, the key goal of the lander was to sniff out organic chemicals. One of the more unusual experiments on Philae was to pick out any complex organic molecules and measure something called chirality. Most organic chemicals can be arranged either in one configuration, or its mirror image.* On Earth, all naturally occurring organic chemicals are 'right-handed', although no one is sure why. Finding out whether organics on comets are right-handed, left-handed or a mixture of both would help make it clear what variety of organics were available at the beginning of the solar system, and help

---

* While you might expect chemicals of different chirality to behave the same way, they absolutely do not. The most tragic example was the drug thalidomide. While one chirality of the chemical helped to alleviate morning sickness, the other caused severe birth defects and miscarriages.

researchers work out why Earth has this disparity. Philae would also be able to take the first direct measurement of the isotopic ratio of water on the comet's nucleus, rather than in the tail or coma. Together, these two results would help answer critical questions about how the ingredients of life first made their way to the surface of Earth.

Alongside all of these, the lander would carry one half of a radio experiment called CONSERT. This would work with the main Rosetta spacecraft, broadcasting radio waves through the comet's nucleus. Just like InSight uses the changes in seismic waves to look inside Mars, CONSERT would use these radio waves to map out the internal structure of 67P.

To do any of that, though, Philae needed to land without bouncing off again. Because the comet has such low gravity, if the lander rebounded from the surface with a speed faster than 1m/s (a slow walking pace) it would escape into space. To prevent that happening, Philae would be released with a relative speed of less than 0.5m/s. Each of the lander's three feet had ice screws to dig into the ice surface, while a harpoon could fire into the ice, securely fixing Philae in place. These would also help keep the lander from being ejected into space if a jet of gas erupted nearby.

After detaching from Rosetta, Philae would have around 60 hours of battery life. This would give it more than enough time to complete its primary science goals, taking measurements from the ice and taking several CONSERT measurements. But space engineers are never ones to settle for 'good enough' and so Philae's solar panels would allow it to recharge and continue monitoring the comet as it passed around the Sun, observing how the surface environment changed.

The mission needed a fair amount of luck and time, not to mention international collaboration, but it would be worth it if they could pull it off. After a decade in the making, on 2 March 2004, Rosetta and Philae were both ready to head off on their mission from ESA's launchpad in French Guiana, in northern South America.

Now began the long wait. Comets travel incredibly fast. For a flyby, this isn't too much of a problem, as you don't need to match their speed, just slow down enough to be able to take photographs. Rosetta, however, would need to keep pace with a comet travelling at 135,000km/h. Building up the speed was going to take more than big rockets. Rosetta had to get three gravity assists from Earth and another one from Mars.

Along the way, Rosetta stopped by an asteroid or two, taking the chance to test out its navigation systems and its remote instruments. By 2011, the spacecraft was looping out into the cold depths of space where it would encounter 67P. With nothing to see and little sunlight, the spacecraft went into hibernation on 8 June to conserve power. For thirty-one months Rosetta slumbered, until 19 January 2014. Rosetta was now 9 million km from the comet and closing. The spacecraft alarm went off on time and it phoned in on cue, much to everyone's relief.

Two months later, in March 2014, Rosetta made its first sighting of 67P. Rosetta spent the next few months getting ever closer, performing a complicated series of manoeuvres until the comet was just 100km away. On 6 August, the Rosetta team announced that the spacecraft was finally at the comet.

The first images back from the comet were a surprise to both the public and scientific teams alike. The comet was a contact binary – two round lobes fused together. This was an unexpected boon to planetary scientists hoping to learn how the early bodies of the solar system clumped together. They now had a front-row view of what two bodies stuck together actually looked like. To the public, however, it was the comet's uncanny resemblance to a rubber duck that launched a thousand Photoshop parodies.

Now in orbit, the spacecraft photographed 67P from every angle. As the comet rotated every 12 hours, almost every part of the surface was illuminated at one time or another. The different light conditions helped reveal the surface texture, a vital quality for the Philae team as they tried to find the perfect place to set down their lander.

As usual, there was a war between wanting to land somewhere interesting and somewhere safe. As well as the usual boulders and crevasses to navigate, there was the ongoing worry that a jet might blast the lander off into space, even after a safe landing.

Eventually, a site named Agilkia (after an island on the River Nile near Philae) was picked as the primary landing site. It was located on the smaller lobe – the duck's head. It was just interesting enough to still be safe, with 6 to 7 hours of light per rotation to charge the solar panels.

On 12 November 2014, the day of the landing arrived. Hundreds of scientists were involved in the creation of the mission, and many of them travelled to the European Space Operations Centre in Darmstadt to witness the launch. ESA had taken a leaf out of NASA's book and had been publicising the event for months, drawing hundreds of reporters. The centre was filled to the rafters with people excitedly waiting to hear news of Philae's successful landing.

Behind the scenes, however, the operations team were having a more troublesome day. While checking out Philae's systems in the run-up to the landing, the team discovered that the gas propulsion system that would prevent the lander from bouncing off into space if it came down too hard wasn't working. Fortunately, the mechanism was only ever intended as a failsafe.

The lander itself had no propulsion, solely relying on Rosetta for aim. The orbiter was able to get in close to the comet, just 22.5km away. As 67P was still relatively inactive, there was little danger of the lander being thrown off by a jet.

At 08:35 a.m. GMT, Philae separated from Rosetta, the orbiter snapping one last farewell image as the tiny lander disappeared into the blackness. It took Philae 7 hours to reach the surface.* As it fell, the lander flipped out its landing legs, preparing for contact. Back on Earth, all they could do was wait for the signal blip indicating the lander had touched down on the surface of the comet. The blip came at 16:03 UTC (Coordinated Universal Time). Cheers erupted around the complex.

'It was a moment of real joy,' said project scientist Matt Taylor, a man as well known for his tattoo-covered limbs (including one of Rosetta on his leg) as his command over the Rosetta mission.** 'Everyone was hugging each other, there was no more rank or political hierarchy or anything else, it was a victory of human kind over a mythical object.'

While the crowd was distracted, crying and hugging each other, the expressions of the landing operations team changed from elation to concern. Something was wrong. The harpoons meant to secure the lander to the surface hadn't deployed. Philae had bounced.

The surface wasn't as soft as expected and without the thrusters, Philae had rebounded, soaring back into the air at a sedate speed of 38cm/s – just below the comet's escape velocity. Philae flew up 1km above 67P, before coming down 2 hours later. It rebounded once more before finally coming to rest.

---

*   With not much to be done but wait during this time, the landing team often referred to this as the '7 hours of boredom', in reference to the '7 minutes of terror' for the Martian landings.

** *Legends of Space*, Episode 10: 'Rosetta and Philae', Euronews, 23 November 2017, www.euronews.com/2017/11/23/legends-of-space-ep-10-rosetta-and-philae

The team now had little idea where the lander was. Philae's first images showed how precarious its position was – it was on its side, barely holding on to the surface with a single ice hook. Worse, Philae was in the shadow of a cliff. Instead of 7 hours of light per rotation, it would have less than 2. Not long enough to recharge. Philae only had its 60-hour battery time left to live.

Not knowing which instruments survived the unexpected bounce, the team got to work, only to find that Philae had got there ahead of them. The lander was programmed to start a 40-minute series of experiments as soon as it landed; however, the sequence started after the first bounce, meaning it had taken several 'surface' readings while 1km above the comet.

Now that it was on the surface to stay, Philae did a quick check out of its main experiments. Out of ten instruments, only two had failed – the APXS and the instrument intended to measure chirality. Both required contact with the comet's surface, but the angle meant Philae just couldn't quite reach.

Not knowing how securely Philae had landed, the team were unsure if drilling would only serve to dislodge Philae's tenuous grasp, making matters even worse. After Philae's working experiments had finished, they decided to drill anyway. Wasn't it better to dare and fail, than play it safe and never know? In the end, Philae extended its drill out to the full 50cm it was capable of, but it was just too far away.

Philae held on for 64 hours before the batteries gave out. The team left the lander to recharge, hoping there might be enough sunshine at least to phone home. But Philae remained silent as the weeks went by. Ground Control continued to listen as the comet got closer to the Sun, in case the increased sunlight allowed it to charge or a jet happened to knock Philae into a better position.

In June 2015, the lander managed to send out the merest hint of a communication – a chirp. Unfortunately, at the time the comet was highly active, and the team didn't want to risk bringing in the Rosetta Orbiter closer for a chance at picking up a signal from a probably useless lander.

In total, the lander made seven communications with Rosetta. On 9 July 2015, Philae made its last call then fell silent. Permanently.

The team had tracked down the landing site to a region known as Abydos. Philae's ultimate final resting place was discovered a few months before the end of the mission when Rosetta got close enough to snap an image of the region. The lander was wedged on its side under an overhang, one leg sticking up in the air, just as they had thought.

Philae wasn't destined to be the only thing to land on comet 67P. After two years following the comet around the Sun and over 8 billion km of travel, Rosetta was finally done. To prevent the spacecraft from drifting off into deep space, potentially causing a problem for a future mission, the team decided to set the spacecraft down on the comet. On 30 September, it became the second object to softly set down on a comet, albeit, the fourth landing.

While Philae might have been less successful than hoped, it was only one small part of a huge mission. As a whole, Rosetta was a colossal success. ESA had followed a comet throughout its lifecycle, sending back so

Philae's final resting place. Rosetta's OSIRIS camera managed to track down the wayward probe after months of searching, taking this picture on 2 September 2016. (ESA/ Rosetta/ MPS/ UPD/ LAM/ IAA/ SSO/ INTA/ UPM/ DASP/ IDA: www.esa.int/Science_Exploration/Space_Science/Rosetta/Philae_found)

much information that scientists are still trying to understand it all as I'm writing this, three years afterwards. For ESA, the mission taught them just as much about how to run a huge space mission as it did about comets.*

Just as with Mars, the real power of the Rosetta mission was the joint work of an orbiter to gain a global image, and a lander to study close up. Here, however, we will concentrate on what Philae taught us.

The bounce meant that Philae managed to get a look at two different sites, albeit only very briefly at one of them. From what it did see, however, the two sites were obviously very different from each other. Sunny Agilkia had a layer of dust over an unexpectedly hard sublayer, which is probably what caused the probe to bounce. Meanwhile, shady Abydos was a rugged terrain of ice cliffs and boulders with little dust at all. The current thinking is that as the comet passes close to the Sun, the volatiles in the top layers sublimate away. Whatever's left behind creates a crunchy layer around the comet's icy heart, which then crumbles to dust.

One instrument that was remarkably successful despite everything was CONSERT. The instrument discovered that 67P is empty headed – 75 to 85 per cent of its head part is empty space. Alongside Deep Impact's results, it seems likely that comets are loosely held together piles of slush.

Although many of the experiments meant to look at the nucleus's composition failed, Philae did manage to find traces of several organic chemicals, although any of the more complicated molecules were either in very low levels, or completely absent. Unfortunately, the two experiments that failed to deploy were those that would have given a deeper understanding of exactly what the comet was made of and what role it played in the creation of our solar system.

The landing mission might not have gone entirely to plan but the combined work of Rosetta and Philae had gifted the world with a complete overview of a comet. The pair had come to a placid rock of ice, then watched it burst to life as it neared the Sun, only to watch it fall back asleep as it passed out once more into the cold depths of space. It was an

---

*   Rosetta Project scientist Matt Taylor has often joked that the biggest thing we learned from Rosetta was that putting 32m-long wings (aka the solar panels) on a comet producing a vigorous 'wind' of outgassing probably wasn't the smartest idea in the world.

epic quest to follow a comet, and one which planetary scientists will look back on for years to come.

These missions revealed the fascinating world of comets to humanity. Each comet was different, but they all appear to be packed with water and the prebiotic chemistry that potentially helped to kick start life on Earth.

One thing comets certainly didn't bring to Earth was water. Rosetta found the deuterium/hydrogen ratio of 67P was three times that of Earth's. After different experiments at different comets, it seemed that comet water just is not the same as Earth water. If comets didn't bring water into the inner solar system, that left the second of the transient space rocks floating around our cosmic neighbourhood – asteroids.

27

ASTEROIDS

It's easy to see how a massive lump of ice otherwise known as a comet could bring water to Earth. Not so obvious is how seemingly dry asteroids might have done so. Despite its arid appearance, there's a lot of water inside an asteroid, only it's locked up inside the rock.

Like comets, asteroids are planetary leftovers. But where comets formed in the cold outer regions, asteroids were created in the inner solar system. As this was much hotter, elements like silicon and iron were molten, meaning they could interact and form far more complicated molecules, before cooling together to form rocks. Over time, these rocks began to clump together, trapping inside some of the more volatile chemicals which were a gas at these temperatures, such as water. Over time, enough rocks clumped together to form a planetoid and eventually planets.

Asteroids are a mixed bag, with dozens of different types. Some are made of pristine material that has never been part of a planet. Others are fragments of planetoids blasted apart when they collided with another space rock flying around the early solar system. If the planetoid was large enough, it might have begun to differentiate – the molten minerals separating out into different layers depending on their density. The precise make-up of these layers is complex, but generally speaking, heavy elements, such as iron, sink to the centre to create a metallic core, while the lighter elements remain in the outer layers. If these planetoids get destroyed, then these layers create a wide array of asteroids. Those that were once the planet's core are almost entirely metal, while others are more like stone.

The different routes to asteroidhood means these space rocks are incredibly varied, with dozens of subcategories based on how much carbon, silicon, or metal they have. It's also why so many people want to study asteroids. They allow a direct look into not just the building blocks of the planets, but the layers normally far below the surface.

There are more self-serving reasons for studying asteroids. They have long been touted as the greatest reservoir of natural resources at our disposal. The water that is still trapped within them could one day be split apart to create rocket fuel, turning asteroids into refuelling stations for exploring the solar system. Meanwhile, other space pioneers are looking for more immediate gains, planning to mine asteroids that are rich in metals such as platinum or gold.*

Perhaps the most self-serving reason is to mitigate asteroid impacts. Every year, dozens of asteroids strike Earth. Most of these are small and break up in the atmosphere, with only a small portion reaching the ground in the form of meteorites. But every few million years Earth is struck by an asteroid large enough to disrupt the entire planet. It's not a question of *if* there will be another big asteroid but *when*. Aware of the danger, several space agencies study asteroids so that if we do find a killer asteroid is on course for Earth, we'll at least have some idea of what to do about it.

Most people only think of asteroids as residing in the belt between Mars and Jupiter but, in truth, they are everywhere. Thousands fly around the inner solar system (a worrying number crossing Earth's orbit), while others are caught up in gravitationally stable points near the larger planets. Wherever you are going in the solar system, the chances are you'll fly past an asteroid. So why not stop by? Most missions to the outer solar system have tacked on at least one asteroid flyby, if for no other reason than to test out their camera equipment.

The first dedicated mission to an asteroid wasn't considered until the 1990s. At the time, NASA was heading into the 'faster, better, cheaper' era, and was hunting for the first potential Discovery mission. As asteroids offered an easy target with a high scientific return and big publicity potential, it wasn't surprising that NASA sent a spacecraft to one of the many near-Earth asteroids.

---

*   There's lots of debate over whether we should be pillaging new worlds. In reality, space mining could help to provide the raw material needed for solar panels, catalytic converters and other environmentally friendly technologies. Currently, these are mined on Earth, often in highly polluting and dangerous ways. In the future, it could be possible to move the most polluting aspects of industry into space, where they cannot hurt Earth's delicate biosphere.

In late 1993, NASA selected the Near-Earth Asteroid Rendezvous (NEAR) mission, later renamed NEAR-Shoemaker in honour of Eugene 'Gene' Shoemaker, a leading asteroid researcher who was tragically killed in a car crash in 1997. NEAR-Shoemaker launched on 17 February 1996, heading for asteroid 433 Eros. Appropriately enough for an asteroid named after a god of love, NEAR-Shoemaker entered into orbit on Valentine's Day 2000, becoming the first human object to orbit an asteroid.

It was a simple spacecraft, with four solar panels extending out like the petals of a flower from a cylindrical gold bus. It spent the next year at Eros, coming within just 24km of the surface. As the first asteroid mission, its goal was simple – find out what asteroids look like up close. The mission focused on imaging the surface, to uncover its composition, size and density.

As the spacecraft reached the end of its mission, the operations team thought about what to do with the spacecraft. The usual way of disposing of a spacecraft was to crash it, but Eros was so small that NEAR-Shoemaker might just bounce back off. The team realised that it might be easier to land on Eros and a daring plan emerged.

The procedure ended up being more like a low-speed docking than a landing, with the final touchdown happening at just 6.5km/h – around a brisk walking pace. During the slow approach, the camera was constantly taking images with increasing detail, with the last from 129m away.

The spacecraft touched down on 12 February 2001, coming to rest on the tips of its solar panels. To everyone's surprise, the spacecraft continued transmitting from the surface. NEAR-Shoemaker had become the first ever asteroid landing mission, and NASA hadn't even done it on purpose.

The spacecraft operated until 28 February 2001, but ultimately succumbed to the -170°C temperature. It now lies silently on Eros, where it will probably remain for billions of years, a testament to what can be achieve when we dare to try the unexpected.

The mission showed that landing on an asteroid was relatively easy, giving mission planners the courage for something more ambitious. Why simply go to an asteroid, when it would be all too easy to bring some of it back? Until then, the only way we'd been able to look at an asteroid up close was examining meteorites, but these are contaminated from their time on Earth's surface. A sample returned from an asteroid would be pristine.

As sample-return missions go, asteroids are definitely one of the most straightforward. They're easy to get to, don't require thrusters, parachutes or harpoons to land on and even the trickiest part of any sample-return mission – relaunching – is relatively simple. One of the reasons that a Mars sample-return mission is taking so long is because relaunching from the surface requires a huge amount of fuel, which either needs to be brought from Earth or created on the Martian surface. On an asteroid, however, the escape velocity is so low that the average person could launch into space by jumping.

With such a tempting target, it wasn't long before someone gave a sample-return mission a go. Rather than one of the old favourites (Europe, NASA or Russia), it was another nation making its mark in space to launch the first flight – Japan.

Japan came to the space stage relatively late. After the Second World War, the occupying Allied forces imposed strict rules on what the nation was allowed to do with rockets. This was meant to curtail the development of missiles but had the unexpected effect of stymieing space exploration. Over time, the rules loosened, and Japan created three agencies to deal with different aspects of space exploration – military, commercial and scientific. One of these was the Institute of Space and Aeronautical Science (ISAS), located at the University of Tokyo.

ISAS's first planetary missions were Sakigake and Suisei to Halley's Comet, followed by a lunar orbiter in 1990 – the first Moon mission since Luna 24 some twenty-five years before. After these successes, the agency looked for a new mission that would make its mark as a new space agency. ISAS had been holding study groups with NASA since the late 1980s and was very aware that it had no hope of competing with the United States directly. What it could do was create a different kind of mission. NASA relied heavily on recycling technology to keep the costs down on their expensive missions. Japan, meanwhile, could come up with something completely new.

The idea they settled on was an asteroid mission called Hayabusa, meaning 'falcon' in Japanese. Like NEAR-Shoemaker, Hayabusa would spend months getting to know the asteroid before landing on the surface. It would then take a sample and return it to Earth.

It was certainly ambitious; potentially too ambitious for a relatively inexperienced nation. But Japan wanted to stand out as a major player. In 1995, Hayabusa was funded.

The spacecraft team chose to visit the tiny asteroid Itokawa, measuring just 500m across. Hayabusa followed the traditional 'gold box with solar panel wings' of most satellites, but with a long tube – known as the sampler horn – jutting out the bottom. Hayabusa would shoot the asteroid with a small projectile travelling at 300m/s, just below the speed of sound. The sampler horn would then hoover up the ejected dust, transfer it to a secure return capsule and start making its way back to Earth. Easy.

The mission also carried a tiny lander, known as MINERVA. The probe was a cylinder just 10cm in height and 12cm in diameter, weighing under 600g. It was loaded up with cameras and thermal sensors. Rather than having to use wheels or tracks to move around, the lander would take advantage of the low surface gravity and hop around the asteroid to spy on several different locations. It was an elegant mission but one that would push the fledgling agency to the limits of its capability. If Japan could pull it off, it would be a major coup.

Hayabusa launched on 9 May 2003 from the Uchinoura Space Center. A few months later, in October 2003, the Japanese Government decided to unite the three space agencies to form the Japanese Aerospace Exploration Agency (JAXA), bringing the scientific and commercial aspects of space-flight under one aegis.

As Hayabusa was in no rush to get to Itokawa, it used an ion engine. The basic mechanics of making a spaceship go is always the same – if you throw something out one side of a spaceship, the spaceship will move the other way. The more you throw out and the harder you throw it, the faster your spaceship goes. Conventional rockets burn fuel in a chemical reaction, heating gas that flies out the back, pushing the spacecraft onwards. It's messy, but it works. The problem is, it requires a lot of fuel to get enough oomph to go anywhere, which makes the rocket heavier, requiring more fuel and meaning that chemical rockets are used sparingly.

Ion thrusters, however, use electric fields to speed up gas to enormous velocity. This only creates a tiny amount of thrust but uses a miniscule amount of fuel, so the engine can be left on throughout the mission. With no air resistance to fight against, the spacecraft slowly accelerates over months or even years. It might take longer to get where it's going, and just as long to slow down, but it's incredibly fuel efficient.*

---

*   Put another way: chemical rockets are kid racers who've seen too many *Fast and Furious* movies, slamming the pedal to the floor to get off the line and power round corners, then coast along the straights. Meanwhile, an ion engine is a middle-aged driver who carefully accelerates away from the light and insists on driving at 56mph on the motorway to get the best fuel economy.

Ion engines weren't new – NASA's Deep Space 1 had used one back in 1998 – but this was the first time Japan had used the technology. Hayabusa still relied on a beefy chemical rocket to get off Earth and head towards the target but once in space the ion engine took over.

However, things were far from plain sailing for the spacecraft – Hayabusa encountered stormy seas. In November 2003, the largest solar flare on record erupted, heading straight for Hayabusa. These colossal explosions throw out charged particles and pockets of magnetism, both of which are extremely hazardous to electronics. The flare struck Hayabusa, injuring its solar panels and reducing the amount of power it could send to the electric ion thrusters. The engine still worked but it was limping – the journey would take longer than expected. Thankfully, the orbit of Itokawa was such that Hayabusa wouldn't miss it entirely.

Hayabusa began its approach to the asteroid in September 2005. It hovered 7km away, following it rather than going into orbit. But it seemed that Hayabusa's problems were rapidly coming to a middle, as two of the three reaction wheels broke. These rapidly spinning wheels help keep spacecraft straight, stopping them from tumbling. A spacecraft can operate just fine with one broken wheel but losing a second means the mission has to rely on chemical thrusters to keep steady and these have a limited supply of fuel.

Struggling to keep its balance, the spacecraft began the surface survey that proceeds any landing attempt, creating a global map of the asteroid. Map done, JAXA chose a landing site and moved Hayabusa to hover 3km from the surface. On 10 November, Hayabusa released a 10cm target marker meant to help its ranging system accurately measure the distance to the surface.

While the target marker hit the astcroid, the same cannot be said of the MINERVA lander. The command to release MINERVA was sent from Earth but there was a 32-minute light delay to Hayabusa. In that half an hour window, the spacecraft realised it was closer to Itokawa than the minimum safe distance and started to back away automatically. Hayabusa's coding had no failsafe in place, and so blindly followed the order to release MINERVA. The diminutive lander floated away into solar orbit, obediently searching for an asteroid surface that wasn't there for 18 hours, when it drifted out of Hayabusa's range, never to be seen again.

There was nothing to be done but concentrate on the landing of the mother spacecraft itself. The first landing attempts started on 19 November. The spacecraft had descended towards the surface, cutting off its engines 17m out as planned and the spacecraft's communications dropped out, as expected. When Hayabusa checked back in, it was 100km from the surface, in safe mode and tumbling slightly. Something had gone wrong.

Over the next few days, the Hayabusa team deciphered the spacecraft's logs to uncover what had gone awry. It seemed Hayabusa had, in fact, touched down. However, during the approach, it spotted an obstacle. It attempted an emergency take-off at an awkward angle. The spacecraft ended up bouncing on the surface twice before coming to rest. The impactor that was meant to kick up dust hadn't fired and so the lander had lain on the surface for half an hour, soaking up heat from the sun-drenched rock. Then the spacecraft got its act together and launched back into space. The entire ordeal confused Hayabusa's delicate programming enormously, forcing it into safe mode until it reconnected with Earth and the safety of home.

The landing team decided to make a second attempt on 25 November. At first, this appeared to have gone to plan. It hadn't. Again, the projectile failed to deploy. It appeared the mechanism had been damaged during the first landing – the spacecraft was meant to deal with the frigid conditions of space, not a long rest on a hot stone.

The two landings had very much not gone to plan but Hayabusa had made contact with an asteroid twice. Even if it wasn't as much as intended, the spacecraft may have collected some asteroid dust. JAXA decided to send the return capsule home, so they could at least see if they'd caught anything.

Of course, Hayabusa's troubles weren't over yet. The beleaguered spacecraft's orientation thrusters developed a fuel leak. While trying to fix the problem, the spacecraft lost its position and communications became spotty. The window to return to Earth in December came and went.

JAXA finally managed to wrestle control back in March 2006, by which point the return capsule had missed the direct connection. It was going to have to take the scenic route. Missing the window added another three years to its odyssey but the spacecraft persevered. On 13 June 2010, the capsule finally returned home, crashing to Earth in the Australian desert.

The mission might have been a disaster but Hayabusa was finally back home. After the trip back to Japan, the scientists cracked open the capsule to see if all the work had been worth it. Had they managed to get any asteroid dust?

The answer was … sort of. Hayabusa had captured less than a milligram of material. Using these few precious grains, geologists were able to determine that Itokawa was a type of asteroid known as an LL chondrite, meaning it was low in iron and metals in general, and Itokawa had once been part of a larger asteroid. However, the biggest thing they learned was how much potential there was if they'd managed to collect a real sample. Sometimes, when it comes to space, you just have to keep trying and refuse to give up.

The main problem with Hayabusa had been pure bad luck. While spacecraft are built to weather solar flares, no one expected Hayabusa to get hit with one of the biggest ever seen. It wasn't long before Japan considered redoing the Hayabusa mission, hoping for better luck.

The agency officially gave the go-ahead for a follow-up, Hayabusa 2, shortly after the first mission's capsule returned to Earth in 2010. It aimed to launch in 2014 and had a relatively modest budget of ¥16.4 billion (around $150 million) as it could reuse most of the hardware from the first mission, building on its successes while correcting its failures.

One tweak was adding a new way to kick up material. As well as the bullet-like impactor Hayabusa had failed to fire, Hayabusa 2 would have a copper weight like the one Deep Impact used, although weighing just 2kg. As Hayabusa 2's impactor would not only be lighter but slower than Deep Impact, it was crammed with explosives to accelerate it into the surface. The explosion would ensure there was plenty of material to pick up, while also exposing deeper layers that hadn't been weathered for billions of years.

The new mission would take another run at landing a hopping probe on the surface. Alongside three MINERVA-II landers would be a fourth lander from Germany named MASCOT (Mobile Asteroid Surface Scout). The cube-shaped lander would investigate the surface's composition and magnetic properties and would also hop around the surface.

The flight spacecraft would have another ion drive and all the cameras and instruments it needed to support the landing. This time, the spacecraft was given an extra antenna so it could communicate with Earth even as it approached the surface. Hopefully, this time, the mission would go to plan.

Hayabusa 2 launched towards the 1km-wide asteroid Ryugu on 3 December 2014. With no mega-flares throwing it off its stride, the spacecraft matched the asteroid's elliptical orbit in three years, beginning its final approach in June 2018. Early images showed the asteroid was a diamond shape, bulging around the middle. The surface was rough and covered in boulders with several deep depressions, although on a global scale it looked fairly uniform.

The spacecraft mapped out the surface for another two months, while the science team hunted down the perfect place to land. The number of boulders was causing some concern. The landing problems from the previous Hayabusa had been down to a rock confusing the sensors and the team didn't want another obstruction to send this landing awry, too.

The initial plan had been to have multiple touchdowns at different spots across Ryugu, but the team struggled to find even one safe place to land and decided to settle for a single location. On 21 September, Hayabusa dropped the first two MINERVA-II landers onto the surface.* Within hours, the first images from the surface of an asteroid were circulating on social media – still with lens flare and distortion – showing a surface strewn with black pebbles.

A few weeks later, it was MASCOT's turn. The hopper touched down on 3 October. Unlike the MINERVAs, which were equipped with solar panels, MASCOT only had 17 hours of battery life to work with. During that short lifespan it took as many images and temperature and magnetic field readings as it could, along with a few spectrometer measurements to gauge the asteroid's composition. After two asteroid days on the surface (it takes Ryugu 7.6 hours to rotate), the rover hopped to a new location, repeating the readings there until the battery died a few hours later.

Then came the main event – Hayabusa 2 setting down on the surface. As the asteroid passed behind the Sun (and out of radio range) in mid-November, the landing didn't take place until early 2019. On 21 February (22 February in Japan) Hayabusa 2 prepared to touch down.

As the assembled media and VIPs prepared to watch the landing, they were taken by surprise when the Mission Control team began cheering a successful landing 35 minutes earlier than expected. There was a nerve-wracking communication loss as the spacecraft pushed its sample horn into the surface, but the spacecraft soon reconnected with Earth. As far as the spacecraft could tell, it had its sample.

Now Hayabusa 2 was ready to make its mark and leave a new crater on the surface of Ryugu. On 5 April, the spacecraft released its explosive package, before quickly flying around the far side of the asteroid to

---

*   Hayabusa 2 was only 60m above the surface during the drop and managed to snap a
      few images of its own shadow on the surface of Ryugu.

dodge the blast. It did leave behind a tiny spacecraft, DCAM3, to watch the demolition. Once the dust settled, Hayabusa 2 returned to survey its handiwork. The blast had created a 20m-wide crater, much bigger than expected, revealing the pristine layers below the comet's surface. On 11 July, the spacecraft performed its second touchdown, in the middle of the asteroid's new crater.

Both touchdowns appeared to have grabbed a sample from the surface, but no one will know for certain until Hayabusa 2 gets back to Earth. The spacecraft headed for home on 15 November 2019 and is due to arrive back in late 2020. Then the real scientific work can begin.

Japan wasn't the only nation inspired by the semi-success of Hayabusa. In the United States, asteroid scientists were studying the mission themselves, planning how NASA could arrange a similar mission to their own asteroid.

In 2009, NASA was selecting its latest New Frontiers missions. The New Frontiers programme was set up in 2002 for mid-level planetary missions, with a budget of $850 million. Amongst the shortlist for 2009 was the Origins, Spectral Interpretation, Resource Identification, Security, Regolith Explorer spacecraft (OSIRIS-Rex). The mission was, broadly speaking, the same as the Hayabusas – fly up to an asteroid, characterise it, take a sample and return it to Earth. However, with over four times the budget of Hayabusa 2, OSIRIS-Rex could afford to be more ambitious. As well as cameras and instruments to map the topography of the asteroid before landing, OSIRIS-Rex would be able to analyse the space rock's composition while still in orbit. The comprehensive picture of the asteroid would give much-needed context when the analysis teams finally got their hands on the samples themselves.

Such detailed mapping takes a long time. While Hayabusa was only at Itokawa for a year and a half, OSIRIS-Rex would spend that long examining the asteroid before even attempting a landing. As the problems of the Hayabusa mission had, in part, been caused by a previously unseen surface obstruction, the NASA team hoped to prevent a similar disaster.

To understand the reason why OSIRIS-Rex was doing all this, it's probably easiest to decode the full name of the mission. The spacecraft would help determine the ORIGINS of the solar system by returning a pristine sample of a carbon-rich asteroid from the early days of planetary formation.

It would use three different spectrometers to measure the global properties of the asteroid. This *in situ* SPECTRAL INTERPRETATION could be compared to measurements from Earth-based observatories, helping to refine future remote observations. This would identify resources across the asteroid which could be useful to future exploration. The spacecraft would also investigate the Yarkovsky effect, where the Sun's heating on a rotating asteroid can cause its orbit to change over time. As this could potentially push an asteroid into a collision course with Earth, risking the SECURITY of the planet, astronomers are keen to study exactly how the effect works. Finally, the spacecraft would EXPLORE the REGOLITH's texture, morphology, geochemistry and spectral properties at the sample site.

In 2011, NASA officially selected OSIRIS-Rex. The mission would launch in 2016, bound for the 500m-diameter Bennu, a near-Earth asteroid named after the heron associated with the Egyptian god Osiris. If all went to plan, the spacecraft would arrive in 2018, returning to Earth five years later in 2023, hopefully with a good sample of asteroid rock on board.

The timeline meant that OSIRIS-Rex and Hayabusa 2 would be operating at the same time, albeit destined for different asteroids. Knowing they could help each other out, the two groups swapped information, team members and worked to learn from each other's triumphs, as well as their mistakes – a far cry from the secretive world of the early Space Race.

Despite OSIRIS-Rex following the traditional boxy appearance of orbiters, it had a strange elegance. Its two solar panels flicked up to resemble wings while the wide cone of the sample-return capsule jutted out at the front like a beak. Beneath the spacecraft, the slender sample horn extended out like a flamingo's leg. Hayabusa may have been the falcon, but it was OSIRIS-Rex that looked like a bird.

On 8 September 2016, OSIRIS-Rex was ready to start its long journey towards Bennu. It would take the spacecraft two years to reach its new home, finally beginning the approach phase in August 2018. At that time, the asteroid was still 2 million km away, little more than a distant dot, but large enough for the cameras to target it and start the long procedure of closing in.

OSIRIS-Rex officially arrived at the asteroid on 3 December 2018, approaching to 7km from the surface – close enough to start characterising the asteroid. In appearance, Bennu was remarkably like Ryugu. The two asteroids were a similar shape and covered in the same rough surface of boulders and dents. The spacecraft used a laser to map out the surface of the asteroid, creating a full 3D model of Bennu.

As 2019 began, the spacecraft reached a stable orbit.* The spacecraft came within a mile of the surface – the closest any spacecraft has come to an object it was studying without actually landing.

It quickly became apparent that OSIRIS-Rex had arrived at an excellent target. After just a week, the spacecraft's spectrometers confirmed the presence of water across the surface. In fact, it looked like there might have been liquid water on the asteroid at some point. As Bennu is too small to have held on to liquid water, that meant it had once been part of a much larger rock.

Like Ryugu, Bennu's surface was far more boulder-ridden than anticipated. This posed a big problem. OSIRIS-Rex planned to get its dust sample using a 'touch-and-go' technique, where the spacecraft flies into the asteroid and places a collector funnel against the ground. The spacecraft then blasts nitrogen gas, causing a cloud of dust to plume up off the surface, which is channelled into several collectors on the spacecraft.

The method is relatively straightforward, but only works if the funnel comes down on a soft, dusty surface. The procedure could deal with some larger grains of dust, but anything too big risked damaging the funnel, while a large pebble could injure the spacecraft itself, preventing it from returning home.

From the preliminary survey of Bennu from Earth, the science team had expected to find several regions where there were no rocks larger than a few centimetres. In reality, though, there were barely any patches without large pebbles. Mission planners found just four regions and even these were much smaller than they'd hoped. The initial mission specifications called for a region that was 25m across with no pebble larger than 2cm, but the largest areas NASA could find were 5–10m in size. It would make the touchdown risky as there was little room for error, but the team had no choice.

The final decision came in December 2019. OSIRIS-Rex would touch down in a region known as Nightingale. The touchdown itself is due to happen in June 2020, after this book has gone to press, so hope-

---

*   At the time, US congress and the White House were in the middle of a feud over funds for President Trump's Mexican border wall. Across the nation, federally funded agencies – like NASA – were forced to shut up shop as they couldn't pay their workers. Fortunately, NASA has a contingency, meaning that ongoing missions were still staffed. OSIRIS-Rex could at least continue working, even if no new plans were made until the shutdown ended.

fully, dear reader, you should have seen many images of rejoicing mission controllers already.

However, it wasn't just the pebble-ridden surface that came as a surprise to the OSIRIS-Rex mission planners – there were pebbles being thrown off it, too. In March 2019, NASA revealed that the spacecraft had spotted plumes of particles being ejected from the asteroid. While some of these fell back down, some were thrown off far enough to escape into space. A few even got caught by the asteroid's gravity and are now in orbit.

This is the first time that jet activity has been seen at an asteroid, so planetary scientists are still investigating what could be causing it. Current theories point towards it being something to do with Bennu being very wet for an asteroid. OSIRIS-Rex will keep a look out for any more plumes. The task shouldn't affect the mission's timeline too much – the planners scheduled in an extra ten months of observation time, just in case they found anything unusual while at the asteroid. Their forethought has paid off.

The find once again reaffirms that the line between comets and asteroids might not be as clear-cut as we used to believe. The discovery of a wet asteroid brings us back to the overriding question: Did the Earth's water come from asteroids?

Although only a few grains were brought back by the Hayabusa spacecraft, a team at Arizona State University were able to analyse the tiny amount of water within them. They discovered that the asteroid contained somewhere between 160 to 510 parts per million of water. While this might not be a lot of water in every asteroid, the Earth has been hit by a huge number of these space rocks over the last 4.5 billion years, and the amount of water adds up over time, especially if there are many asteroids as wet as Bennu.

However, the real kicker came when they measured the isotopic content. It was almost identical to that found on Earth. It is at least possible that asteroids played some role in bringing water to our planet. We'll just have to wait until Hayabusa 2 and OSIRIS-Rex come home to find out how big a role.

For future missions, most space agencies aren't as concerned with which asteroids might have hit Earth in the past but which ones might pay an unwelcome visit in the future. While there are currently several telescopes surveying the skies to track down potential killer asteroids, little work has been done on what would happen if we actually found one. The Asteroid Impact and Deflection Assessment (AIDA), a joint mission between NASA and ESA, hopes to change that in around 2022.

Despite what Bruce Willis might have taught you in *Armageddon*, nuking an asteroid just as it's about to hit Earth won't stop you getting hit. Even if you could completely blow an asteroid apart, which is unlikely, the pieces will still be heading towards Earth. It would be like trading being hit by a cannon ball for being shot with a shotgun.

Most plans to avoid being wiped out involve slowing down an asteroid so that by the time it reaches Earth orbit, the Earth has already moved on and the asteroid misses.* The easiest way to slow down an asteroid is to fire something heavy into it. Of course, doing this requires a precision impact. If the impact goes wrong, then the asteroid might not get nudged enough or be pushed in the wrong direction. It seems foolhardy to wait until a killer asteroid is bearing down on us to start testing the technology, hence AIDA.

AIDA is made up of two separate missions to the asteroid Didymos, a double system with a 775m and 165m asteroid orbiting around each other. The basic principal is that NASA will slam into the smaller asteroid, known as Didymoon, with the Double Asteroid Redirection Test (DART) spacecraft – essentially just a big weight with an engine on the back. Then, astronomers will look to see what that does to Didymoon's orbit, both on the ground and *in situ*, when ESA's Hera mission arrives a few years later to have a look. Hopefully, no big bad asteroid will have appeared in our back garden before it's had the chance.

---

*   Of course, this relies on knowing about an asteroid several years ahead of time. Both NASA and ESA run searches for potentially hazardous asteroids and now think they've found around 99 per cent of the civilisation-killing asteroids, those over 1km. None seems to be a threat any time soon. The numbers aren't quite as good for the city killers – asteroids above 140m in size. They only know where about a third of those are. In case you needed something else to worry about today.

28

ICY MOONS

For much of this book, we've looked in our near neighbourhood – the inner solar system. Be it the planets and moons that live there permanently or the asteroids and comets that swing by from time to time. These are the most visited locations in our cosmic neighbourhood for the practical reason that they are close and easy to reach. But that isn't to say there aren't a whole host of other worlds out there that planetary scientists dream of landing on.

While the planets of the outer solar system are gas or ice giants, without a solid surface to land on, there have been missions that plunged through their thick atmospheres. Most of these have been orbital missions reaching the end of their life, but one Jupiter spacecraft, Galileo, had a specific atmospheric probe. The spacecraft dropped the probe on 7 December 1995, lasting 57 minutes before being crushed by the pressure. In that time, it travelled around 600km through the atmosphere, suspended beneath a parachute. The probe was buffeted by 500km/h winds, suffered both extreme cold and heat and was shaken by turbulence as it passed through the atmosphere's layers.

However, our exploration of the planets is, largely, driven by the self-centred need to find another world like Earth. Partly, this is because these robotic explorers are proxies for our own exploration, so there is a selfish desire to only explore the worlds that humans might one day be able to walk on. Another part is that the questions that drive space missions have almost all ultimately come down to understanding our own planet and humanity's place on it. It's not really a surprise that the highest-priority targets in the outer solar system are those most like Earth.

Although the gas giants are very much not Earth-like, they host hundreds of moons. Many of these are little more than rough chunks of ice and rock a few kilometres across, but several are large enough to resemble planets themselves. In fact, two of them – Jupiter's Ganymede and Saturn's Titan – are larger than Mercury.

These worlds are varied in more than just size. Jupiter's closest moon, Io, is the most volcanically active world in the solar system, as the pull of the giant planet's gravity moves the internal rocks around, heating them up. But then, the next moon out, Europa, is a frozen world that hides an ocean under its surface and has huge jets of water bursting through its crust. Go out to the very edge of the main planets, to Neptune's moon, Triton, and you'll find a world where the water acts like rock, with slushy 'magma' flowing from cryovolcanoes made of ice.

There are several other worlds that show signs of a layer of liquid water deep down under the surface. Just like Triton and Europa, Saturn's Enceladus has water jets erupting from it, feeding into the planet's rings. Jupiter's moon, Ganymede, wobbles on its axis, a sign that a liquid layer is sloshing about inside the world.

As we've seen in previous chapters, the mantra of searching for life beyond Earth is to 'follow the water', and so these moons have long attracted the interest of astrobiologists. However, until recently, there have been few missions to these outer worlds.

Exploring them is incredibly difficult. They are far away, making them hard not just to reach but to communicate with. They are far from the Sun, meaning that solar panels have to be huge to generate enough power, if they are not outright inadequate. Around Jupiter, there's the added problem of its intense magnetic field, creating bands of radiation that fry any electronics that come too close. Even if a mission did manage to land on an icy moon, a probe would need a 20km-long drill to even get near the interesting stuff.

Most of the missions that have been to the outer solar system have been flybys, with only three reaching orbit – Galileo and Juno at Jupiter, Cassini at Saturn. With such difficulty even orbiting these gas giants, it's no surprise that only one mission has even attempted to put down a lander on an icy moon, the Saturnian moon, Titan.

Among the many worlds in the solar system, Titan undeniably stands out – it is simultaneously much like Earth and yet also completely alien. Telescope observations as far back as 1944 showed it has an atmosphere, albeit made of methane, giving rise to hopes of finding life on the moon. When Pioneer 11 flew past Saturn in 1979, it saw this atmosphere was as thick as Earth's, making it the only moon known to have a significant atmosphere.

In November 1980, Voyager 1's attempt to photograph the surface was scuppered by a thick orange haze, although it did measure the temperature at around -179°C; cold enough for ice to be as hard as rock, with methane and ethane pooling like water does on Earth. When Hubble took infrared images of Titan, it showed bright and dark regions – some of them the size of Australia.

As the Sun's ultraviolet light breaks down methane, something on Titan has to be renewing the gas. Could there be ice volcanoes on the surface, erupting methane like the volcanoes of Earth pump out carbon dioxide? Or perhaps there was some exotic hydrocarbon cycle, like the water cycle that drives the rains of our own planet.

Titan has long enthralled geologists – an alien landscape of ice mountains drenched by ethane rains that ran into Earth-like liquid oceans under a thick, orange sky. It was only a matter of time before someone planned a visit.

NASA began seriously considering the mission in 1977. In 1982, a European and US working group came up with the idea of a joint mission to Saturn. One nation would supply an orbiter to go around the giant planet, the other a lander to drop on Titan. The first visit to what was truly an alien world, the lander would focus on learning as much about both planet and atmosphere as possible. As the lander descended to the surface, it would analyse the density and composition of the air, while also taking photographs of the landscape, giving clues to how geology worked on a world made not of rock, but ice.

Originally, ESA wanted to build the orbiter, but at that point ESA had only staged one planetary mission, Giotto. It seemed foolhardy to trust the new agency to build such a major spacecraft, so they switched roles. NASA would build the orbiter, Cassini, while ESA would work on the lander, Huygens.*

NASA planned to build Cassini alongside the CRAF comet investigator, but Cassini and its lander quickly became more complicated than anyone had anticipated. Cassini alone was shaping up to be a billion-dollar

---

*   Giovanni Domenico Cassini was a seventeenth-century Italian astronomer who discovered four of Saturn's moons and realised that the ring around Saturn had several divisions. Christiaan Huygens was the Dutch astronomer who discovered Titan.

mission. In 1992, NASA was forced to choose between CRAF and Cassini. Losing Cassini would mean leaving ESA's lander without a ride, so CRAF met its end.

While NASA built the orbiter, ESA worked on the lander. Huygens would descend into the unknown. Scientists knew the atmosphere was thick but not *how* thick, making designing a parachute incredibly difficult. Even if Huygens did make it safely to the surface, it could just as easily splash down in an ethane sea as slam into solid ice. Like the early Venus probes, Huygens would have to deal with all eventualities. Unlike the Venus probes, there would be no Huygens 2 if everything went wrong.

Given the number of uncertainties, it wasn't considered a mission-critical goal to land in one piece, although obviously the design team tried their hardest to make it possible. They ensured Huygens would float but also gave it enough padding that it could survive a hard landing.

The first part of the descent would be controlled by a heat shield* until it had bled away enough speed for the parachutes to take over. Here, the design team hit a snag. The atmospheric scientists wanted as much time falling through the upper layers of atmosphere as possible, slowly travelling downwards so they could examine the changing composition and density of these layers. This required a big parachute. However, the spacecraft would be running on battery power with a limited lifespan. If Huygens was to have any time on the surface, then the descent time need to be kept under 2.5 hours. In the lower, denser layers of the atmosphere, such a big parachute would severely slow the spacecraft's fall, cutting into the battery time. So, Huygens borrowed a trick from the Venera probes – two parachutes. A large one would slow the initial fall, then cut away to deploy a smaller one so the spacecraft could reach the surface quickly.

On board the probe itself would be instruments to measure the atmospheric composition, pressure and temperature, as well as several cameras. During descent, the spacecraft would slowly spin, panning their view around to capture the whole surface.

To get all this information home, Huygens would transmit to the Cassini Orbiter across two radio channels, which would then relay it back

---

* The heatshield was a logistical nightmare. The French, who were building them, had developed the technology for missiles and weren't keen on handing over the details of secret military hardware for testing. ESA managed to work it out, but it does demonstrate why so many international space missions never get off the drawing board.

to Earth. All of this was set inside a metal container that resembled two pie tins, one flipped on top of the other; albeit for pies measuring 1.3m across.

By early 1997, both halves of the mission were built, tested and on their way to Florida to launch on board a Titan IV, a rocket usually reserved for military use. Everything was going well until the spacecraft was installed on top of the launch vehicle.

Saturn is 1.5 billion km from the Sun, too far for solar panels to be any use, so instead Cassini used radioactive thermal generators (RTGs) for power. As well as electricity, RTGs produce a huge amount of heat, which is great for keeping warm in the cold of space and less good when you're trying to stay cool in the Florida heat. Radioactive material can't just be turned off, so while it was still on Earth the spacecraft had to be constantly cooled.

While Cassini was being installed on top of the Titan IV, the air-conditioning unit was accidentally set far too high. The intense blast of air shredded Huygens' insulation, potentially firing tiny pieces of debris into the delicate workings of the probe. With the launch window rapidly approaching, the entire probe was unmounted, carefully unwrapped and the damage fixed.

With all that palaver, Cassini missed the first week of its launch window. Fortunately, there was still plenty of time. As the usual space enthusiasts gathered to watch the eventual launch on 13 October, they were joined by a less enthusiastic group – twenty-seven protesters. For several months, activists had been trying to get the probe blocked so as to prevent sending nuclear material into space. If the launch went awry, there was a risk that plutonium would rain down on the local populace – a concern backed by some of NASA's own scientists. Although the agency had tested the RTG to ensure it was as indestructible as possible, these tests were, as former NASA Emergency Preparedness Officer Alan Kohn put it, based on a '"SWAG" – scientific wild-assed guess'.*

The majority of NASA's staff were confident of its safety, though. Beverly Cook, who had overseen the building of the RTGs, pointed out, 'I'll have my family at the launch and I certainly wouldn't do that if I thought I was putting them in any danger.'**

However, the protesters served as a reminder that despite the furore surrounding a mission launch, spaceflight isn't always universally loved. The

*   L. Weiss, 'Cassini Controversy', motherjones.com, 30 September 1997.
**   A. Lawler, 'Cassini Faces Last-Minute Hurdles', *Science*, Vol. 277, 12 September 1997.

group were arrested, and the launch went ahead – with neither the rocket nor RTG coming to any harm.

The delay would ultimately prove beneficial. The most efficient time to launch is almost always in the middle of the window, when Cassini actually managed to make it off the ground. Mission planners usually choose to launch early, on a less-efficient path, rather than wait and risk being cancelled due to bad weather or someone turning the air conditioner up too high. The shorter flight time required less fuel, leaving more for use at Saturn.

It took Cassini seven years to reach Saturn. While most of this time was spent in hibernation, the team ran systems checks along the way. One of the checks that the communications team wanted to run was during an Earth flyby in 1999 that would give the probe a gravitational boost. They aimed to mimic the signal between Huygens and Cassini during descent. When the results came back, they discovered the signal coming in was complete gibberish. If this happened at Titan, the lander wouldn't be able to communicate its findings back to Earth and the mission would be useless.

The communications team quickly tracked down the issue. Cassini's motion away from Earth was shifting the frequency of the signal. It was only a few parts per million awry but the range of Cassini's radio was set too narrow. This range was built into the spacecraft's hardware so there was no way to change it from Earth. 'We have a technical term for what went wrong,' said John Zarnecki, leader of the surface lander team. 'It's called a cock-up!'*

Thankfully, there was a way around the potential disaster. Originally Cassini was going to drop Huygens during the first flyby of Titan. If it was dropped on the third pass instead, the change in approach speed and angle meant the relative speed between Cassini and Huygens was small enough for the radio to cope with. The new plan had the added advantage that Cassini would get a few glimpses at the moon before attempting the landing.

The team would turn the radio's heater on a few hours early to heat up, which would also shift the frequency slightly. It would eat into the battery life but a short mission that gets data is better than a long mission that gives you none.

---

*   D.M. Harland, *Cassini at Saturn: Huygens Results* (Springer, 2007) p.231.

After a seven-year flight, Cassini arrived at Saturn on 1 July 2004. The spacecraft got its first look at Titan from over several hundred kilometres away before managing two closer flybys over the next month. Cassini's radar system pierced through the haze, exposing the moon's landscape while its spectrometers began to look for complex molecules in the atmosphere.

The third flyby would be the big one. On Christmas Eve 2004,* Huygens separated from Cassini. It would be another twenty days before the probe reached the moon. For three weeks the lander fell towards the surface, the battery only being used to keep the alarm clock ticking. On 14 January 2005, at 04:41 UTC, that alarm went off and Huygens woke up.

Throughout the approach, Earth was watching. Almost every telescope that could see Titan at the time was lending its eyes to the mission. Hubble was supposed to be among them, but in a stroke of bad luck the instrument the team needed failed just a few months before landing.

The atmosphere in Mission Control was tense. Europe's only previous landing attempt, the British Beagle 2, had been a failure. Even though that hadn't been an ESA-led endeavour, its spectre hung over the Titan mission as the communications team waited to hear from the lander, 67 light minutes away. Their ability to control Huygens ended when it separated from Cassini, and the probe was now operating on its pre-programmed code. Whatever was going to happen during the landing was already locked in place.

The probe began to feel the first tug of drag from the atmosphere around 1,300km out. As the probe fell, and the atmosphere grew thicker, the drag built and with it the heat. Temperatures reached up to 10,000K, hot enough that the probe would have glowed white hot. Although several telescopes looked for the 'meteor trail' left behind by Huygens, none managed to make it out.

Several of the radio telescopes that were trained on the probe picked up the radio signal coming in from Huygens, starting with the Green Bank Telescope in West Virginia, United States. The signal was far too weak to decipher but it was there. As the science team wouldn't receive the actual signal until several hours after the landing, when Cassini turned back to

---

*   Space missions have a habit of timing their big days for public holidays. The Cassini crew at JPL overseeing the release were at least allowed to don Santa hats to mark the occasions, and the traditional lucky peanuts came in the form of festively coloured peanut M&Ms.

Earth to begin the transfer, the faint radio blip was reassuring; at least Huygens wouldn't disappear without trace as Beagle 2 had done.

When Huygens reached the right speed, it popped its main parachute. After 15 minutes, this cut away and the smaller chute deployed for the probe's rapid descent. Turbulence meant that Huygens suffered a bumpy ride but around 10km from the surface, the air calmed, and the probe fell more smoothly during its final approach. 2 hours and 27 minutes after the descent began, Huygens touched down on a solid surface* with a speed slightly less than 5m/s. So much for the romantic idea of sailing on Saturnian seas.

Cassini heard signals from the surface for over an hour after the landing until the orbiter disappeared over the horizon around 70 minutes after touchdown. Back on Earth, radio telescopes picked up the faint signal for another few hours but when Cassini itself came back in range, the probe had fallen silent, its batteries now dead.

A few hours after the encounter, Cassini swung around and transmitted its data back to Earth. At first, the signal came in, but it was carrying no information. After 5 minutes of panic, the data began to come in clearly. They'd done it! They had a signal back from Titan.

It quickly became apparent that all was not well from the surface. Of the radio's two channels, only Channel B was carrying any data. Channel A was nothing but zeros. The problem turned out to be gallingly simple. When Cassini had been commanded to listen out for Huygens signal, a single line of code meant that the radio picking up Channel A hadn't been turned on. A stupid mistake that was too simple to be picked up during testing.

While the most critical information was transferred over both channels as a precaution against something just like this, several teams had chosen to split their data between the two, effectively doubling their potential scientific output. Over 350 images were lost, and one instrument – the Doppler Wind Experiment – required the much more accurate clock that Channel A was equipped with but Channel B lacked. 'I have never felt such exhilarating highs and dispiriting lows than those I experienced when Green

---

*   The lander team held a sweepstake on exactly what time the lander would touch down, which was won by Principal Investigator John Zarnecki. Some declared shenanigans, although Zarnecki maintains that his only advantage was having a full overview of the mission. His prize was a bottle of malt whisky, which he cracked open at 2:30 a.m. to share with his team.

Bank reported the signal indicating "all is well", only to discover there was no signal on A-channel!' said the experiment's leader, Michael Bonn.[*]

The team began to stitch together what images they did have, but couldn't make them match up until someone realised that the probe was spinning the wrong way. No one was sure how it happened. The probe was ejected from Cassini with an anticlockwise spin of 7.5 revolutions per minute. At some point, it had switched direction. When the landing team reviewed the results of the test drops, they realised that the probe had changed direction then as well. No one had thought to check which way the probe was spinning – only that it was. Fortunately, the misaligned spin had very little effect on the mission.

The pasted-together panoramas showed a mountainous landscape of craggy rocks, split apart with vein-like structures that looked suspiciously like the drainage channels of lakes.

During the descent, the probe had been buffeted by winds. At high altitude, 120km up, these winds blew as fast as 120m/s but then calmed to a much more sedate 1m/s near the surface. Looking at the images of the moon, however, it seemed the wind must occasionally gust stronger because at least some of the surface features had been shaped by the wind.

One of Huygens' instruments that looked for aerosols – tiny particles suspended in the moon's atmosphere – found that there is a steady fall of 'organic snow', made up of amino and nitrile groups, all of which are building blocks for life on Earth. The 'snow' seems to have smoothed out some of Titan's rougher edges, implying it's a widespread phenomenon and raising hopes that this steady snowfall could at least pave the way for some primitive form of life on the planet.

The atmosphere, like Earth's, was predominantly made up of nitrogen – around 90 per cent. This is high compared to the amount of carbon we've seen on Titan. If the moon was created with the same levels of nitrogen and carbon as we think existed in the early solar system, it suggests there's carbon locked up in the planet – possibly in the form of methane below the surface. This would track with the idea of cryovolcanoes slowly releasing methane into the atmosphere. The idea is backed up even further by the presence of Argon-40 on the surface, which is used as a way of tracking potential volcanism.

[*]   Harland, *Cassini at Saturn: Huygens Results*, p.289.

The team knew that the lander had come down on a solid surface with the consistency of wet sand, but the lander's instruments indicated that an obstruction gave way under the lander's weight shortly after landing. There were suggestions that the landscape was coated in a hard crust with a soft layer beneath – something like crème brûlée. Equally, the lander might have come down on a pebble, which crushed after a moment.

It wasn't hard to see how the lander might have hit a rock. The place was littered with pebbles a few centimetres across and apparently made from hard ice. Most excitingly, though, they looked like rocks that had been smoothed by a flowing river. True, in this case the river was made of methane, but they were another piece of evidence for a liquid flow across Titan's surface. Along with the other instruments on Huygens, it seemed likely the moon's methane was indeed acting like water does on Earth. In short: the rain on Titan is mainly methane.

While the main ESA team were working out what kind of world they had landed on, another group were frantically trying to reclaim some of the Channel A data that had been thought lost – the Doppler Wind Experiment. Cassini might have only listened to Huygens for 70 or so minutes, but there were lots of telescopes on Earth that had listened for much longer. One of them, the Parkes telescope in Australia, was still listening in as ESA discovered the Channel A data was missing from Cassini's download. When Parkes looked at their own feed, they could see that Channel A was transmitting fine, even if Cassini couldn't hear. With Saturn about to set in Australia, a series of frantic calls whipped around the European stations, who all hurriedly pointed their telescopes towards the distant moon, with some engineers climbing into theirs to reconfigure them on the fly, hoping to recover something.

Ultimately, the real rescue came from Parkes joining forces with Mopra, another Australian telescope 500km away. The Mopra hard drives were delivered to Parkes on a chartered flight and over a frantic day of number crunching and telecommuting with a researcher from Finland, the team managed to do months' worth of data analysis in one day. At 6 a.m., the Parkes team announced they'd found a signal from the experiment. Not only could they recover the lost Doppler Wind Experiment data, they could use it to pinpoint the location of the lander on the surface. It was a suitably Hollywood finale to what was an epic mission.

Despite what movies would have you believe, most scientific enterprises don't happen overnight. Knowing the data would take months to analyse, ESA originally only released three images – one from the surface, two from the air – saying that the rest of the catalogue would be released a week later when they had a better idea of what they were looking at. However, to the picture-hungry media, alarms started ringing. Had something gone wrong with the images? Were these three the only ones they had?

Some enterprising soul found that although the agency had only released three images, the entire catalogue was available if you just changed the number on the end of the URL. Within minutes, the full Huygens gallery was all over the internet as amateurs and enthusiasts began to stitch them together to create their own beautiful panoramas of icy mountain ranges.*

At first, the team at the University of Arizona who were responsible for the leak were horrified. They'd accidentally let slip scientific data before it had gone to the official science team. Now, anyone could scoop the Cassini-Huygens science teams who'd been patiently waiting for years. As the months rolled by, though, no one pipped them to the post. The images were spectacular, but without the calibration data and software to understand them, they were scientifically meaningless.

The leak was more a sign of ESA's naivety as a space agency, underestimating just how voracious the media and public alike are for pictures and instant results. They learned from their mistakes, and by the time the Rosetta mission rolled around the ESA media machine rivalled that of NASA.

Cassini itself spent thirteen years orbiting Saturn – far beyond even the wildest hopes of the initial mission planners – flying past Titan over 100 times. In 2017, the spacecraft reached the end of its life.

NASA couldn't just leave Cassini in orbit, as there was a risk it could crash into one of the ice moons. As well as Titan, plumes of water have been seen coming from Enceladus, so either of them could be potential havens for life. If Cassini came down on one of these, not only had it not been subjected to the strict planetary-protection cleaning of Huygens and other landers, it had a red-hot RTG that could potentially melt through the ice crust, carrying that contamination to the liquid ocean below. After

---

*   In most cases, panoramas created by amateurs are more alluring than the science releases. While ESA have to maintain the scientific integrity of the image, people at home can blur out the edges and amp up colouration to produce beautiful, if not entirely accurate, landscapes.

a series of flights between Saturn and its rings, Cassini was crashed into the gas giant.

Cassini had been a monumental mission, and Huygens was far more successful than anyone on the team had ever dared hope. Despite the many problems of both orbiter and probe, the team landed on an alien world that lies ten times further from the Sun than Earth does. In doing so, they had revealed a world with ice mountains and methane rains.

The short-lived Huygens mission only served to wet NASA's appetite for Titan. In 2019, the agency announced they are returning. This time, rather than attempting to land on the surface or float on the sea, the agency aims to skim through the air with a drone-like spacecraft known as Dragonfly. Taking advantage of Titan's thick atmosphere, Dragonfly will hop from place to place, traversing kilometres at a time to examine all over the moon, searching for signs of prebiotic chemistry that hint at the moon's habitability. Could, or perhaps even has, life evolved on the surface of Titan? Dragonfly hopes to find out. The mission aims to launch in 2026, arriving at Titan in 2034.

Titan isn't the only icy moon getting attention. If it's possible to land on Titan, is it also possible to land on Europa, Enceladus or even the distant Triton orbiting Neptune?

For many of these moons, the real place of interest isn't the icy surface, but what lies beneath – the subsurface oceans. Several innovative designs have been put forward for underwater explorers that could be sent to Europa or Enceladus, including a robotic squid designed by Cornell University that uses the moon's magnetic field to generate power. But, before such a mission can even be considered, NASA would need to work out how to create a robotic spacecraft capable of boring through 20km of ice to reach the ocean. This might be feasible if a human were manning the drill. However, as much as we might want to send an astronaut to set foot on one of these icy worlds, the technology to keep people alive at several hundred degrees below zero, billions of kilometres from Earth is still a long way off.

There might be another way to taste these oceans, though. A Stardust-like spacecraft could fly through one of the water plumes seen on Enceladus and Europa, bringing home the waters of a distant moon for us to find out just how alien these alien worlds really are.

PART 5
FORWARD TO
THE FUTURE

29

BACK TO THE MOON

In recent years, the drive to explore the solar system has gone through a shift – instead of forging ever further outwards, people are returning their gaze towards home. This shift, once again, is down to human space exploration with at least three different nations set on putting another set of boot prints on the Moon – China, India and the United States.

Since 2 November 2000, there's been a constant human presence in low-Earth orbit, courtesy of the International Space Station (ISS). Astronauts living and working on the station have been teaching mankind not just how to survive in space but how to thrive there. Several nations now think they have learned enough to start venturing further into the solar system. For the first time since the Apollo landings, space agencies are actively planning to take a human to the Moon. This time, they intend to leave behind more than just flags and footprints. This time, they mean to stay.

Although it will be years, if not decades, until a human is ready to set foot on the Moon, this new focus has spurred a renaissance for robotic missions to the lunar surface. Partly, these robots will scout the way, just as Surveyor did. As well as testing the landscape to see if it could physically support a base, they would hunt out possible resources, such as water locked up in the rock or investigating whether the regolith can be turned into something like concrete.

The last landing mission to the Moon had been Luna 24 in 1976. Space robotics have moved on a lot since then and there is a lot still to learn about our nearest neighbour. It will also be necessary to learn how to land on the Moon affordably and reliably. Any sustained lunar presence will have to be supported by robotic missions, either sending them beforehand to aid in setting up a permanent base or to bring supplies once astronauts settle in.

One nation that is already on its way is one that, until now, has not featured much in the annals of planetary exploration – China. The Chinese space programme began in the Cold War days of the 1950s. Initially, the nation worked with its Communist allies, the Soviet Union, but the relationship between the two states soured when Khrushchev came to power. Lacking the political drive to compete in the Space Race, China's progress soon fell behind that of the United States and the Soviet Union, but it did still progress.

In 1967, Chairman Mao decreed that China should begin work towards putting a *taikonaut* (the Chinese word for astronaut) into space. Internal political turmoil set these plans back for many decades, but in 2003 China finally launched Yang Liwei into orbit, thereby becoming the third nation capable of launching people into space.

The next long-term goal was to get a taikonaut onto the surface of the Moon. First, the Chinese would have to learn how to work in space. Initially, China planned to join the ISS, but in 2011 the United States enacted a law banning NASA from working with China,* and therefore effectively banning them from the ISS. Instead, the nation has been forced to build their own space station, the first parts of which they're aiming to launch in the coming years.

In the meantime, China has mounted the Chang'e programme, a series of robotic missions to the lunar surface named after the Chinese goddess of the Moon. Unlike the early United States' and Soviet lunar programmes, Chang'e isn't racing anyone to be the first to explore the Moon. Instead, the Chinese National Space Administration (CNSA) are free to take their time and do things properly with a systematic three-stage plan – orbiting, landing and sample return.

Chang'e 1 and 2 fulfilled the first stage, orbiting in 2007 and 2010 respectively, surveying the Moon down to a resolution of just 1m. After several years of mapping the Moon, it was time to move on to the second phase of the project, landing, with Chang'e 3.

---

* The reason behind the ban is often cited as fear of the links between China's space programme and its military. However, several key figures involved with the decision were also outspoken on China's poor treatment of its Christian citizens.

On 14 December 2013, Chang'e 3 successfully landed in the Mare Imbrium (Sea of Showers) on the lunar near side. It was the first time China had soft landed anywhere and the first lunar landing in thirty-seven years.

Like the Soviet Union, China adopted a policy of secrecy, so not much was known about the mission in advance, although it seemed that Chinese aerospace engineers drew inspiration from landers of the past. In terms of what the lander would do, though, Chang'e 3 was a different animal to what had gone before. While previous missions had had their heads down in the lunar dirt, Chang'e 3 would have one eye upwards, towards the stars.

The Moon has long been touted as a potential base for astronomical observatories. It has no atmosphere, which on Earth soaks up several very useful wavelengths of light such as infrared and ultraviolet. Chang'e 3 had several astronomical experiments, including an ultraviolet telescope looking at distant galaxies and bright stars, while another ultraviolet camera looked at the energetic plasma around Earth. When on the surface, the camera revealed the glowing halo of plasma, while the ultraviolet telescope managed to snap as many as 10,000 images every month. To the best of Western knowledge, the lander and ultraviolet telescope were still working over six years after landing.

One part of Chang'e 3 would be looking downwards, however – its rover. As usual, although the static lander has done much worthy of applause, it's the charismatic rover that got the headlines. It even had an adorable name – Yutu, meaning Jade Rabbit after the lunar goddess's companion.[*]

The rover resembles a gold box on six wheels with a stereoscopic camera poking out of the top, bringing it to a final height of 1.5m, with solar panels jutting out on either side. Once again, the rover's design borrowed from its ancestors, taking its wheel design from the Soviet Lunokhod Rovers.

Yutu's aim was to scout the surface for resources To do this, the rover had a robotic arm, allowing it to use an APXS to measure the elemental composition of the Moon's soil at the surface, while a radar would look beneath it. These radar measurements would help to gauge how strong the landscape is, and how well it would be able to bear a large structure, either on the surface or cut into the rock.

Yutu drove off the lander onto the lunar surface on 14 December 2013. Its goal was to explore the Moon over three lunar days (each of which is

---

[*] The myth stems from the fact that the lunar maria resemble the shape of a rabbit standing on its hind legs, the Chinese equivalent of 'the Man in the Moon'.

twenty-nine and a half Earth days), which meant surviving three lunar nights. During the first lunar day, the Chinese Government revealed several images taken by the rover and lander of each other, showing Yutu was happily trundling around the surface. Ten Earth days after landing, the rover powered down and prepared to endure its first long night on the Moon.

The rover reawakened from sleep mode on 11 January 2014 and began making its first inspection of the lunar regolith. The rover was meant to repeat these measurements in several different locations. However, this wasn't to be. On 25 January 2014, the Chinese media announced the rover was having a 'mechanical control abnormality'. The instruments were working, Yutu was still transmitting but the rover wasn't moving. The landing site was home to far more big rocks than the Chang'e team were expecting. Yutu had struck one of these, wounding the rover and rendering it unable to move.

Despite the paralysis, the rover continued to observe its surrounding area. But with every passing night, the instruments' performance waned. Yutu made its last communication in August 2016.

Although its journey was shorter than expected, the rover did uncover that the lander had come down on a lava flood plain, just as they would have expected in a lunar mare. Meanwhile, the lander continued to monitor the sky and communicate with Earth and, as of late 2019, was still apparently calling home.

While Chang'e 3's mobility problems might have stopped it doing everything it set out to do, it achieved its main goal – showing that CNSA could land on the Moon. The agency just needed to work out the kinks of running a rover over the rough terrain. With a soft landing already under their belt, it was time for China to attempt something no other nation had managed yet – landing on the far side of the Moon.

The lunar far side has long fascinated planetary geologists as it has a completely different landscape to the near side, with none of the familiar maria. Gravitational maps of the Moon show the difference is more than skin deep, as the crust of the far side is thicker. Geologists have been waiting decades for someone to send a mission to the far surface, and now finally one was being sent.

The reason for the long wait is communications. Talking to the lunar far side is difficult because the main bulk of the Moon gets in the way, blocking out radio waves, meaning that flight teams have to use a relay station to bounce the signal from Earth to the lander. In May 2019, CNSA launched

Queqiao, meaning Magpie Bridge,* to act as the go-between for not just China's own far-side operations but any other nations who wished to travel there as well. While China might be secretive about its own exploits, the nation was more than happy to configure their hardware to allow for outside collaboration.

Chang'e 4's lander was the same as Chang'e 3, although the rover design was updated to be more reliable, based on what the agency had learned from their first sojourn on the surface. Unfortunately, these updates meant the loss of two key components – the robotic arm and the APXS. Without these, the rover wouldn't be able to make any detailed measurements of the mineralogy of the Moon.

The news was broken to the world's geologists at the Lunar and Planetary Science Conference. 'I think it's not too strong a statement to say the scientists present were dismayed by the thought of a lander being sent to the [region] without such an instrument,' said Emily Lakdawalla, one of the assembled science writers.** They were finally getting a mission to the lunar far side but it would be half-blind.

One thing the rover could look for was water. Chang'e 4 was heading towards the South Pole Aitken Basin, a 2,500km-wide crater left behind by an ancient impact. Orbital investigation hinted at hidden stocks of water, locked up in the lunar soil in the shadows of craters where sunlight never reaches. One of the rover's key goals would be to measure exactly how much water the crater held.

The measurement is of great interest to human spaceflight planners hoping to exploit it as a potential resource for future lunar colonists. The basin has long been touted as a potential site for a lunar base attached to a radio telescope. The far side is the perfect place for such a facility for the same reason it's so hard to communicate with – the Moon blocks out all radio waves from Earth. To test its suitability, the Chang'e 4 lander would have a small radio telescope.

---

* The name is based on the Chinese myth around two of the brightest stars in the sky: Niulang (Altair in the West) and Zhinu (Vega) are ill-fated lovers who are forced to live on either side of a river (the Milky Way). On the seventh night of the seventh moon each year, a flock of magpies take pity on the couple and create a bridge, allowing them to reunite for one night.

** E. Lakdawalla, 'Plans for China's farside Chang'e 4 lander science mission taking shape', The Planetary Society, 22 June 2016: www.planetary.org/blogs/emily-lakdawalla/2016/06220913-plans-for-change4.html

Despite the loss of its arm, Yutu-2 would still be able to investigate some physical properties of the basin. The crater is around 12km deep, meaning it has excavated the deeper levels of the lunar crust, and might even have revealed the mantle. Just like Gale Crater on Mars that Curiosity is exploring, the South Pole Aitken Basin offers a chance to get underneath the Moon's skin without having to drill kilometres into the surface. The rover would once again carry a radar system, allowing it to look deeper still.

Chang'e 4 began its journey to the Moon on 8 December 2018. Although it reached lunar orbit a few days later, the spacecraft waited almost a month before landing. Due to the issues in communicating with the far side, the landing would require complete autonomy. The team wanted to make sure they knew everything they could about the landing site before attempting to touch down on the surface.

On 3 January 2019, Chang'e 4 touched down on the lunar far side in the Von Kármán Crater. Soon enough the rover rolled on to the surface and began exploring. For the most part, Yutu-2 mimicked the previous mission, although it had to operate with a greater level of autonomy due to the communication issues. Thankfully, the lander had come down in a much smoother area than its predecessor, decreasing its odds of suffering the same fate.

The first surprise from the mission, to the wider world at least, was the announcement on 16 January 2019 that a cotton seed on board the lander had sprouted, making it the first plant grown on another world. Chang'e 4 carried a mini-biosphere experiment containing several seeds, fruit fly eggs and yeast, along with the water, air and nutrients they need to grow. The experiment used the natural light on the lunar surface to see if it was possible to grow crops on the Moon. For cotton at least, the answer is yes!

The experiment was short-lived, as it wasn't protected from the chill of lunar night. By the time the announcement was made, the Sun had already set, dropping the temperature to a chilly -190°C, killing all the specimens. Only the cotton seeds sprouted, but even that limited success is a sign that it's at least possible to grow crops that could sustain a future colony on the Moon.

Fortunately, the rest of the rover survived the night. Yutu-2 resumed work and survived well beyond the three lunar nights specified in the mission parameters.

On the rover's eighth lunar day, in July and August 2019, it discovered a strangely coloured 'gel-like substance' that seemed to glisten in the sunlight, which made its operators stop by for a closer look. Using the few instruments Yutu-2 did have, geologists from the Chinese Academy of Sciences realised it was actually a glossy rock, created by the heat of a meteor impact. Despite the rover's lack of instruments, it is still managing to give some geological insight.

Chang'e 4 was a success, and China are now gearing up to launch the next phase of their programme, a sample-return mission. Initially, Chang'e 5 had been designated as a back-up if the first two landings failed, but with the success of both, the rover was removed from the design and replaced with a drill and a return probe. Once off the Moon, the return capsule will dock with an orbiter, then return home with its precious load of moondust.

Chang'e 5 was originally scheduled to launch in mid-2017, but the mission was delayed due to issues with the Long March 5 rocket needed to launch it. The rocket was meant to take over as China's heavy-launch vehicle in 2016, but the first flight failed to reach its intended orbit, while the second fell into the ocean shortly after take-off. The rockets were taken out of use while the problems were fixed. The rocket successfully returned to service on 27 December 2019, so hopefully we should get the first piece of moonrock since Luna 24 before too long.

China is not the only rising space nation using the Moon to set itself up as a major player. Just a few months after Chang'e 4's touchdown, India followed suit with its own lander.

Like most countries, India's space programme has its roots in the dawn of the atomic age and the work of a celebrated scientist. Born into a family of industrialists, Vikram Sarabhai was passionate about making his nation one of the world's best. He set up institutes to advance India's industry through both research and management, as well as helping to bring nuclear power to India.

For our story, however, it was Sarabhai's role in setting up the nation's space agency that is important. Sarabhai saw the benefits that satellite communications could give to the nation's rural communities. It would allow India to show itself off as a technologically advanced country. In 1969, the nation set up the Indian Space Research Organisation (ISRO).

Before long, India had created its own rockets and was launching other people's spacecraft. With much cheaper rates than other launch providers, India carved itself out a niche as a provider of affordable spaceflight. India's space programme has very much been led by practical concerns, and most of their launches have been communications and GPS satellites, although several scientific missions have snuck in.

The first of these were all astrophysical observatories, looking beyond the bounds of our solar system and even our galaxy. Towards the end of the twentieth century, the nation was considering their first foray into planetary missions. The first of these would be a series of lunar missions named Chandrayaan, meaning 'mooncraft' in Sanskrit. Just like the Chang'e missions, these would start with an orbiter, then a lander and rover, ending with a sample-return mission. However, the missions cost a fraction of what other nations were spending.

The first spacecraft was Chandrayaan-1, a lunar orbiter that launched on 22 October 2008. The total estimated cost of the mission was just 3.8 billion rupees ($83 million), well under what NASA would have paid.*

One of the main goals of the mission was to gain direct evidence of water on the Moon, which hadn't yet been confirmed. The spacecraft carried an impactor called the Moon Impact Probe. The probe was a 34kg box that would drop onto the lunar surface, taking pictures as it approached, much like NASA's pre-Apollo Ranger probes. The Moon Impact Probe began its descent on 14 November 2008, impacting 25 minutes later. During descent, a spectrometer showed significant evidence of water in the Moon's southern polar region, findings that were backed up later by other instruments on the orbiter.

With one successful landing on the surface, the nation prepared for a second, although this time they were aiming for a soft landing and a rover. Since 2007, ISRO had been developing Chandrayaan-2 alongside Roscosmos, finalising the design in 2009 with an eye to a 2015 launch. However, the mission was postponed due to a problem several million kilometres away on Mars.

In 2012, the Russians once again mounted a mission to try to land on the Martian moon, Phobos, this time alongside the Chinese. It was to be a sample-return mission, drawing on the designs of the late Luna missions, which were over thirty years old.** The mission failed to even make it

---

*   Part of this was achieved through collaborating with international partners, meaning the probe flew with instruments from the United States and Europe.

**   Russia has never been one to design a new spacecraft for the sake of it. The Soyuz crew capsule, which Russia uses to this day, was first flown back in 1968. It's had a few updates in that time, but the base design remains the same, a testament to the wisdom of 'if it ain't broke, don't fix it'.

out of Earth orbit. No one had checked to see if several key components could withstand the radiation of space.

It turns out they couldn't, and a lack of ground testing meant the issues weren't discovered until the spacecraft was in orbit. In the wake of this embarrassing failure, Roscosmos pulled out of Chandrayaan-2. India proceeded with their lunar mission alone.

The mission finally made it to the launch pad on 22 July 2019, reaching lunar orbit a month later. It carried the Vikram Lander, named after the scientist who had championed India's space agency fifty years before.

On 7 September 2019, the Vikram team gathered in the control room for the landing. The room was an explosion of colour as the many female members of staff had come wearing bright saris to mark the occasion.* As the descent began, everything seemed to be going well. Across the control room, screens showed the green line of the spacecraft's trajectory as it made its way towards the surface. For 11 minutes, the green line followed the red one marking out the lander's ideal path. For 11 minutes the spacecraft was on track. Then, 11 minutes into the manoeuvre, as Vikram was due to tilt itself slightly so that its landing cameras could see the surface, the green line began to veer away. Slightly at first, as it would if the automated hazard avoidance systems were avoiding some obstacle, and then rapidly.

The spacecraft fell silent. The landing team had no idea what was wrong. Was there just a communications failure and some erroneous trajectory data, meaning the lander was sitting on the surface waiting to call home? Or had something gone seriously wrong? The Chandrayaan team continued to wait and listen, hoping their lander would phone home, but heard nothing.

Weeks later, the truth emerged. As the thrusters fired to tilt the lander, things went haywire and Vikram flipped upside down. The spacecraft tried to correct itself, but with the thrusters meant to slow it down now pointing the wrong way, it accelerated towards the surface instead, and crashed. An investigation is currently ongoing into the reason behind the crash, but the finger of blame is currently pointing towards either the landing thrusters, which were a new design, or an oversight in programming. Perhaps a combination of the two.

---

* The Chandrayaan 2 mission was a landmark for women in India's space programme, as it was headed up by two women: Project Director Vanitha Muthayya and Mission Director Ritu Karidhal.

The crash was a setback but won't deter India from its long-term goals. The current head of ISRO, K. Sivan has already pledged to return to the Moon to 'set things right', learning from the mistakes made this time around. Space is hard, especially in the beginning, and this was India's first attempt at a soft landing on another world. Like China, India's long-term goals are human spaceflight and a crewed Moon landing. India has only been working towards sending vyomanauts (*vyoma* means space in Sanskrit) into space on their own rockets since 2007 and hope to make their first launch in 2022.

With the United States, China and India all vying to send people to the Moon in the next few decades it looks like a new Space Race could be heating up. This time, however, the race isn't a sprint, but a marathon – it doesn't matter when you finish as long as you cross the line at the end. That isn't to say that robotic exploration of the Moon has once again been relegated to support human missions. Instead, the baton of robotic scientific exploration has been passed to another group of people – private enterprise.

The move started with the X-Prize Foundation, a non-profit organisation that sets extravagant public competitions to drive technological development. In 2007, they set out the Google Lunar X-Prize, which challenged a privately funded team to land on the Moon by 2014. To claim the prize, the challengers would need to land on the surface, travel at least 500m across it and transmit pictures back to Earth. The winner would take home $30 million, much less than such a mission was likely to cost. The real prize would be the prestige as the company went forward to offer up their lander services to the wider world, as the contest's goal was to kickstart commercial travel to the Moon.*

After the contest was announced, dozens of groups around the world formed to meet the challenge. Some were private companies intending to begin couriering paying customers' payloads to the Moon. Others were

---

*   The X-Prize Foundation had already shown the viability of this plan with the Ansari X-Prize, which asked teams to build a reusable suborbital spacecraft. The winner of that contest, SpaceShipOne, became the basis of the craft that Virgin Galactic plan to use to carry astrotourists into space.

disparate collections of scientists and space enthusiasts working together over the internet.

For all the teams, gathering funding quickly proved a more difficult task than anyone had anticipated. Many dropped out and the competition deadline was pushed back several times, before finally being cancelled on 31 March 2018 without a winner.

Even without the prospect of a prize, several teams carried on building their landers. In 2019, the first team made it to the launch pad. SpaceIL was an Israeli group of former engineers and aerospace workers who had managed to gather $100 million from various sources and donors. On 22 February 2019, their lander, Beresheet (meaning Genesis or 'in the beginning'), launched from Cape Canaveral on top of a SpaceX rocket.

Beresheet went in for landing on 11 April 2019, but 150m from the surface communications cut out and it crashed. Within two days, the company announced that the accident hadn't put them off and were already making plans to launch Beresheet 2.0 within the next few years. The mission might have crashed, but they'd still managed to touch the Moon.

And therein lay a big problem.

The lander was a technological demonstration, so it hadn't carried a big payload. One thing it did carry was a time capsule containing several genetic samples and a few freeze-dried tardigrades.

Tardigrades, otherwise known as water bears or moss piglets, are microscopic organisms that have survived even after being dehydrated, subjected to extreme radiation and exposed to the vacuum of space. When Beresheet crashed, there was a very real possibility the archive had cracked, spilling tardigrades all over the Moon. While the idea of moss piglets running around the lunar surface might sound cute, the idea that a spacecraft had contaminated the Moon with a form of life that could be reanimated horrified planetary scientists.

To make matters worse, no official entity even knew that the tardigrades were on Beresheet in the first place. Currently, there is no international oversight for what can be launched into space other than a treaty banning weapons of mass destruction. Instead, the nation a spacecraft launches from is responsible. When Beresheet launched, the Arch Mission Foundation neglected to mention the organic samples.

In terms of contamination, the crash is probably not that catastrophic. Beresheet was not the first time people have knowingly left living organisms on the Moon. Chang'e 4 took fruit fly eggs and yeast. Even the

Apollo landings left behind sacks of bacteria on the surface in the form of their faecal waste disposal bags.*

Water-bear-gate highlighted that legislation is doing a woeful job at keeping up with space exploration. As not just robots but humans seek to return to the Moon, both with private companies and government agencies, this failure to keep pace could begin to become a problem – one that might make itself known within the coming decade.

Although India and China have stated a long-term goal of one day landing a human on the Moon, there is one nation that is getting very close to actually doing so – the United States. Their plan is to build a space station in orbit around the Moon, called the Lunar Gateway, which will act as a staging hub for human missions to and from the surface.

The timeline for this was moved up in December 2017, when President Trump stepped up the challenge to land the first woman on the Moon by 2024,** a project named after Apollo's sister, Artemis. The Democratic-held House of Congress is refusing to supply the extra funding to make this happen, however, and so meeting this new deadline is looking unlikely.

NASA are still working towards the idealistic 2024, though, and have announced a new landing mission to the Moon, the Volatiles Investigating Polar Exploration Rover (VIPER), which they aim to send to the lunar south pole by December 2022. This will hunt out resources such as water at the poles, much as Yutu-2 has done.

If all goes to plan, things could be about to get very busy on the lunar surface.

---

*   Sacks of poo.
**   This would coincide with the final year of Trump's term in office, should he be elected for a second term. Whether hubris or a love of exploration motivated this decision, I leave for you to decide.

30

SO WHAT HAPPENS NEXT?

The story of space exploration is not finished. There are already several agencies – both government and private – that are preparing to take their next steps out into the wilderness of space. For the first time since the days of Apollo, these steps are not being taken by robots, but humans.

The re-emergence of human spaceflight inevitably raises one question: Should we send humans to space at all when we can send a robot?

Robots undoubtedly have their downsides. Although AI has come on in leaps and bounds in recent years, allowing Curiosity's navigation systems a degree of finesse that was undreamed of fifty years ago, it would still be a stretch to claim that any spacecraft can think for itself. If something goes wrong, they can't work out how to get themselves out of it. They can't decide 'this rock looks particularly interesting, so let's investigate it'.

And what they can do, they have to do monumentally slowly. Curiosity's top speed on flat, hard ground is 140 m/h. Even the most heavily suited astronaut could easily outpace it. Each Apollo landing mission managed to collect 100 times more lunar material than every single robotic sample-return mission put together. And yet ... human missions are exponentially more expensive, risk lives and are limited to where you can feasibly keep a human alive. You can't swap a human's flesh and bones for more radiation-resistant versions, and there's only so far that protective shielding can go. Plus, people expect you to bring human astronauts back. If a space agency wants to avoid a sternly written letter from the UN, they need to work out how to take off from a planet as well as land on it.

Perhaps, then, the future of space exploration isn't a question of pitting humans against robots but getting them to work in unison. Any long-term exploration by humans will likely rely heavily on robots, whether it be in scouting the way as Ranger and Surveyor did for Apollo or laying down the infrastructure needed to support a long-term crewed mission.

It could be that the two work together more directly. ESA is actively investigating telerobotics – remotely operating a robot from a distance. The idea is to send an astronaut, not to the surface of a planet, but into orbit. Once there, they could drop a robot on the surface and operate it in real time, using touch feedback and virtual reality to make the experience far more straightforward than it had been for the Lunokhod drivers. They trialled the idea in 2016, when UK astronaut Tim Peake drove a rover across the surface of 'Mars' while he was on board the ISS. It could be in the future that, while robots are the hands operating in the solar system's most inhospitable environment, it is still human brains that are in control.

Whatever aspects of the actual physical exploration of planets is undertaken by robots, though, there will always be a human element to the mission. It is humans, after all, who look to the heavens and are driven by that primal need to know what is out there. To explore. To learn. Mankind has looked to the stars for millennia and longed to travel to them, and with the talent of engineers, scientists, seamstresses, administrators, politicians, campaigners and a thousand other skilled workers, we have managed to craft ourselves a fleet of robotic ambassadors to do so for us.

And there is still so much left to explore.

ACKNOWLEDGEMENTS

I came up with the idea for this book several years ago when I was looking for a basic guide to planetary landers and rovers and realised it didn't exist, but it has taken many people to make it a reality.

From the folks at The History Press, I would like to thank my commissioning editors: Chrissy McMorris for tracking me down to see if I had a book idea and getting me started on this crazy journey; and Simon Wright, whose invaluable suggestions made this book so much better than it was. Also, a massive thank you to all the other people at The History Press who transformed my words into the book you see today.

Thank you to Declan Waters, who was the first person to read this thing – you gave me the confidence I needed that I hadn't wasted a year of my life writing utter twaddle.

Finally, I would like to send thanks to the friends and family who have supported me. To my Mum, who instilled me with a love of the written word. To my Dad, brother and nieces, for always being more enthusiastic about my career than me. And of course, to my Sam, who has kept me something resembling sane – I could not have done this without you.

SELECT BIBLIOGRAPHY AND SOURCES

Bell, A.J., *Planetary Landers and Entry Probes* (Cambridge: Cambridge University Press, 2007).

Brown, R., et al., *Titan from Cassini–Huygens* (New York: Springer, 2009).

Brzezinski, M., *Red Moon Rising* (London: Bloomsbury Publishing, 2007).

Burbine, T.H., *Asteroids: Astronomical and Geological Bodies* (Cambridge: Cambridge University Press, 2017).

Chaikin, A., *A Passion for Mars* (New York: H.N. Abrams Inc., 2008).

Clark, J.D., *Ignition! An Informal History of Rocket Propellants* (New Brunswick; New Jersey: Rutgers University Press, 1972).

Ezell, E.C., and L.N. Ezell, *On Mars: Exploration of the Red Planet 1958–1978* (Washington DC: Scientific and Technical Information Branch, 1984).

Fimmel, R.O., *Pioneering Venus: A Planet Unveiled* (Washington DC: NASA Ames Research Center, 1995).

Godwin, R., *Surveyor Lunar Exploration Program: The NASA Mission Reports* (Ontario: Apogee Books Publications, 2006).

Hall, R.C., *Lunar Impacts: The History of Ranger Missions* (Washington DC: Scientific and Technical Information Office, 1977).

Harland, D.M., *Cassini at Saturn: Huygens Results* (New York: Springer, 2007).

Harland, D.M., *Paving the Way for Apollo 11* (New York: Praxis, 2009).

Harland, D.M., *Robotic Exploration of the Solar System Part I: The Golden Age 1957–1982* (Chichester: Praxis, 2007).

Harland, D.M., and R.D. Lorenz, *Space Systems Failures: Disasters and Rescues of Satellites, Rockets and Space Probes* (Chichester: Praxis, 2007).

Harvey, B., *Russian Planetary Space Exploration* (Chichester: Praxis, 2007).

Harvey, B., *Soviet and Russian Lunar Exploration* (Chichester: Praxis, 2007).

Kassel, S., *Lunokhod-1: Soviet Lunar Surface Vehicle* (Santa Monica, California: The Rand Corporation, 1971).

Lambright, W.H., *Why Mars: NASA and the Politics of Space Exploration* (Baltimore, Maryland: Johns Hopkins University Press, 2014).

Lindroos, M., 'The Soviet Manned Lunar Program', fas.org/spp/eprint/lindroos_moon1.htm

Lorenz, R., and J. Mitton, *Titan Unveiled* (Princeton: Princeton University Press, 2008).

Lunar and Planetary Programs Division, *Surveyor Program Results* (Washington DC: Scientific and Technical Information Division, 1969).

Manning, R., and W.L. Simon, *Mars Rover Curiosity: An Inside Account from Curiosity's Chief Engineer* (Washington DC: Smithsonian Books, 2014).

Meltzer, M., *When Biospheres Collide: A History of NASA's Planetary Protection Programs* (Washington DC: NASA, 2001).

Mishkin, A., *Sojourner: An Insider's View of the Mars Pathfinder Mission* (New York: Berkley Books, 2003).

Moltenbrey, M., *Dawn of Small Worlds: Dwarf Planets, Asteroids, Comets* (New York: Springer, 2016).

Mutch, T.A., *The Martian Landscape* (Washington DC: Scientific and Technical Information Branch, 1978).

Paquette, J.A., 'The D/H Ratio in Cometary Dust Measured by Rosetta/COSIMA', 49th Lunar and Planetary Science Conference, 2018.

Russell, C.T., *Deep Impact Mission: Looking Beneath the Surface of a Cometary Nucleus* (Pasadena, California: Springer, 2005).

Shepard, M.K., *Asteroids: Relics of an Ancient Time* (Cambridge: Cambridge University Press, 2015).

Siddiqi, A.A., *Beyond Earth: A Chronicle of Deep Space Exploration, 1958–2016* (Washington DC: NASA Office of Communications, 2018).

Spilker, L.J., *Passage to a Ringed World: The Cassini–Huygens Mission to Saturn and Titan* (NASA, 1997)

Squyres, S., *Roving Mars: Spirit, Opportunity, and the Exploration of the Red Planet* (New York: Hyperion, 2005).

Starkey, N., *Catching Stardust* (Bloomsbury Sigma, 2018).

Stooke, P.J., *The International Atlas of Mars Exploration: The First Five Decades* (Cambridge: Cambridge University Press, 2016).

Stooke, P.J., *The International Atlas of Mars Exploration: From Spirit to Curiosity* (Cambridge: Cambridge University Press, 2016).

Wilhelms, D.E., *To a Rocky Moon: A Geologist's History of Lunar Exploration* (Tucson, Arizona: The University of Arizona Press, 1993).

Zakharov, A.V., 'Close Encounters with Phobos', *Sky and Telescope*, July 1988.

Zubrin, R., *The Case for Mars* (New York: Free Press, 1996).

Online Sources

Astronautix: www.astronautix.com

China National Space Administration: www.cnsa.gov.cn/english/index.html

European Space Agency: www.esa.int

Hayabusa 2 Project: www.hayabusa2.jaxa.jp/en

Indian Space Research Organisation: www.isro.gov.in

Institute of Space and Astronautical Science: www.isas.jaxa.jp/en

Jet Propulsion Laboratory: www.jpl.nasa.gov

Messier, D.M., 'Soviet Venus Missions', astro.if.ufrgs.br/solar/sovvenus.htm

Mitchell, D.P., blog – Venus, Soviet space history, computer graphics, science, etc:
    mentallandscape.com

NASA History: history.nasa.gov

NASA Science Mars Exploration Program: mars.nasa.gov

NASA Science Solar System Exploration: solarsystem.nasa.gov

NASA Space Science Data Coordinated Archive: nssdc.gsfc.nasa.gov

Smithsonian National Air and Space Museum: airandspace.si.edu

OSIRIS-Rex: www.asteroidmission.org

The Planetary Society: www.planetary.org

The Space Review: www.thespacereview.com

INDEX